Matrix and Determinant

Mathematical Engineering, Manufacturing, and Management Sciences

Series Editor: Mangey Ram, Professor, Assistant Dean (International Affairs), Department of Mathematics, Graphic Era University, Dehradun, India

The aim of this new book series is to publish the research studies and articles that bring up the latest development and research applied to mathematics and its applications in the manufacturing and management sciences areas. Mathematical tool and techniques are the strength of engineering sciences. They form the common foundation of all novel disciplines as engineering evolves and develops. The series will include a comprehensive range of applied mathematics and its application in engineering areas such as optimization techniques, mathematical modelling and simulation, stochastic processes and systems engineering, safety-critical system performance, system safety, system security, high assurance software architecture and design, mathematical modelling in environmental safety sciences, finite element methods, differential equations, reliability engineering, etc.

Total Quality Management (TQM)
Principles, Methods, and Applications
Sunil Luthra, Dixit Garg, Ashish Aggarwal, and Sachin K. Mangla

Recent Advancements in Graph Theory
Edited by N. P. Shrimali and Nita H. Shah

Mathematical Modeling and Computation of Real-Time Problems: An Interdisciplinary Approach
Edited by Rakhee Kulshrestha, Chandra Shekhar, Madhu Jain, & Srinivas R. Chakravarthy

Circular Economy for the Management of Operations
Edited by Anil Kumar, Jose Arturo Garza-Reyes, and Syed Abdul Rehman Khan

Partial Differential Equations: An Introduction
Nita H. Shah and Mrudul Y. Jani

Linear Transformation
Examples and Solutions
Nita H. Shah and Urmila B. Chaudhari

Matrix and Determinant
Fundamentals and Applications
Nita H. Shah and Foram A. Thakkar

Non-Linear Programming
A Basic Introduction
Nita H. Shah and Poonam Prakash Mishra

For more information about this series, please visit: https://www.routledge.com/Mathematical-Engineering-Manufacturing-and-Management-Sciences/book-series/CRCMEMMS

Matrix and Determinant

Fundamentals and Applications

Nita H. Shah
Foram A. Thakkar

CRC Press
Taylor & Francis Group
Boca Raton London New York

CRC Press is an imprint of the
Taylor & Francis Group, an **informa** business

First edition published 2021
by CRC Press
6000 Broken Sound Parkway NW, Suite 300, Boca Raton, FL 33487-2742

and by CRC Press
2 Park Square, Milton Park, Abingdon, Oxon, OX14 4RN

Library of Congress Cataloging-in-Publication Data

Names: Shah, Nita H., author. | Thakkar, Foram A., author.
Title: Matrix and determinant : fundamentals and applications / Nita H. Shah, Foram A. Thakkar.
Description: First edition. | Boca Raton : CRC Press, 2021. | Series: Mathematical engineering, manufacturing, and management sciences | Includes bibliographical references and index.
Identifiers: LCCN 2020038825 (print) | LCCN 2020038826 (ebook) | ISBN 9780367613167 (hardback) | ISBN 9781003105169 (ebook)
Subjects: LCSH: Matrices. | Determinants.
Classification: LCC QA191 .S47 2021 (print) | LCC QA191 (ebook) | DDC 512.9/43--dc23
LC record available at https://lccn.loc.gov/2020038825
LC ebook record available at https://lccn.loc.gov/2020038826

ISBN: 978-0-367-61316-7 (hbk)
ISBN: 978-1-003-10516-9 (ebk)

Typeset in Times
by Deanta Global Publishing Services, Chennai, India

Contents

Preface

In recent years, matrix and determinant have become an essential part of the mathematical background required by mathematicians, faculties, engineers, researchers and many more. It reflects the importance and wide applications of the subject. This book aims to present an introduction to matrix and determinant, which will be found helpful to the readers. It has been made more flexible to provide a useful content and to stimulate interest in the subject. Every chapter begins with clear statements of definitions, principles, and notations along with illustrative examples to exemplify and amplify the theory and to recall the basic principles which are important for effective learning. Unsolved questions at the end of the first three chapters serve as a complete review of the material. We thank the entire team of CRC Press (A Taylor & Francis Company) for giving us this innovative opportunity to write this book as well as for their unfailing cooperation.

About the Authors

- **Prof (Dr) Nita H. Shah**

  Prof Nita received her PhD in Statistics from Gujarat University in 1994. Since February 1990 she has been the HOD of the Department of Mathematics at Gujarat University, India. She is also a post-doctoral visiting research fellow of the University of New Brunswick, Canada.

  Prof. Nita's research interests include inventory modelling in the supply chain, robotic modelling, mathematical modelling of infectious diseases, image processing, dynamical systems and its applications. She has published 13 monographs, 5 textbooks, and 475+ peer-reviewed research papers. Four edited books, with Dr Mandeep Mittal as the co-editor, have been published by IGI Global and Springer. Her papers have been published in high-impact journals of Elsevier, Inderscience, and Taylor & Francis and she has authored 14 books. According to Google Scholar, the total number of citations is over 3057 and the maximum number of citations for a single paper is over 159. The H-index is 24 up to September 2020, and i-10 index is 77. She has guided 28 PhD students and 15 MPhil students till now and is mentoring an additional 8 students pursuing research for their PhD degree. She has delivered presentations to audiences in the United States, Singapore, Canada, South Africa, Malaysia and Indonesia. Prof Nita is Vice-President of the Operational Research Society of India and a council member of the Indian Mathematical Society.

- **Dr Foram A. Thakkar**

  Dr Foram earned her PhD degree in Mathematics in October 2018 from the Department of Mathematics, Gujarat University, Ahmedabad. Her research area includes mathematical modelling for various social and health issues prevailing in today's world. She has more than 30 research publications in well-reputed international journals, with good indexing and impact factor, and has presented eight research papers at various national and international conferences. Dr Foram served as an Assistant Professor at the Department of Mathematics, Marwadi University, Rajkot, Gujarat, India, from December 2018 to June 2019. She has also reviewed two research articles for the book *Mathematical Models of Infectious Diseases and Social Issues*, edited by Nita H. Shah and Mandeep Mittal.

1 Matrices

1.1 INTRODUCTION

Matrix provides a clear and concise notation for the formulation and solution of variables which are assumed to be related by a set of linear equations. The concept of determinant is based on matrix. Here, we shall first understand a matrix.

A set of mn numbers (or other mathematical objects), arranged in a rectangular array (formation or table) consisting of m rows and n columns and enclosed within a square bracket $[\ ]$, is called an $m \times n$ *matrix* (read as 'm by n matrix').

This $m \times n$ matrix is expressed as $A = \begin{bmatrix} a_{11} & a_{12} & \cdots & a_{1n} \\ a_{21} & a_{22} & \cdots & a_{2n} \\ \vdots & \vdots & \cdots & \vdots \\ a_{m1} & a_{m2} & \cdots & a_{mn} \end{bmatrix}$, where a_{ij} is the

element of the ith row and j^{th} column of the matrix. Hence, the matrix A is sometimes represented in a simplified form by (a_{ij}), $[a_{ij}]$ or by $\{a_{ij}\}$, i.e. $A = (a_{ij})$, $A = [a_{ij}]$ or $A = \{a_{ij}\}$. Usually matrices are denoted by capital letters A, B, C, etc., and their elements by small letters a, b, c, etc.

Note that if the elements of a matrix are real numbers, then it is called a *real matrix*, and if the elements of a matrix are complex numbers, then it is called a *complex matrix*.

The ordered pair having the first component as the number of rows and the second component as the number of columns is called the *order or dimension of matrix*. In general, if there are m rows and n columns of a matrix, then its order is written as (m, n) or (m × n).

For example, $\begin{bmatrix} 1 & 3 & 5 \\ 2 & 4 & 6 \end{bmatrix}$, $\begin{bmatrix} a \\ b \\ c \end{bmatrix}$ and $\begin{bmatrix} 1 & -1 \\ 2 & 3 \end{bmatrix}$ are matrices of order (2×3),

(3×1) and (2×2), respectively.

1.2 TYPES OF MATRICES

1. Row Matrix and Column Matrix

A matrix of the type $(1 \times n)$, i.e. consisting of a single row, is called a *row matrix* or *row vector*, whereas a matrix of the type $(m \times 1)$, i.e. consisting of a single column, is called a *column matrix* or *column vector*.

For example, $\begin{bmatrix} 8 & 4 & 6 \end{bmatrix}$ is a row matrix and $\begin{bmatrix} 1 \\ 2 \\ 5 \end{bmatrix}$ is a column matrix.

2. Zero or Null Matrix

A matrix in which each element is '0' (zero) is called a *zero or null matrix*. A zero matrix is denoted by the symbol 'O'. It is separate from the real number 0.

For example, $O = \begin{bmatrix} 0 & 0 & 0 \\ 0 & 0 & 0 \end{bmatrix}$ is a zero matrix of order 2.

3. Square Matrix

A matrix having the same number of rows and columns is called a *square matrix*. Hence, if m = n, then it is a square matrix. If a matrix A is of order m × n, then it can be written as $A_{m \times n}$. In this case, a square matrix of order n × n is simply written as A_n.

For example, $\begin{bmatrix} 8 & -6 \\ 9 & 4 \end{bmatrix}$ and $\begin{bmatrix} 1 & 0 & 6 \\ 9 & 8 & -2 \\ 3 & 5 & 4 \end{bmatrix}$ are square matrices of orders

2 and 3, respectively.

Remark: The elements $a_{11}, a_{22}, a_{33}, \ldots, a_{nn}$ of a square matrix A_n are called *diagonal elements*. These are also known as the *main or principal or leading diagonal*.

For example, if $A = \begin{bmatrix} 1 & 6 & -9 \\ 2 & 5 & 3 \\ 8 & 6 & 0 \end{bmatrix}$, then the principal diagonal or the

diagonal elements of the given matrix A are 1, 5 and 0.

- *Special Cases of Square Matrix*
  - (a) Diagonal Matrix

    A square matrix in which all the elements are zero except those in the diagonal elements (principal diagonal) is called a *diagonal matrix*. Some of the elements of the main (principal) diagonal may be zero but not all.

    For example, $\begin{bmatrix} 1 & 0 \\ 0 & 2 \end{bmatrix}$ and $\begin{bmatrix} 1 & 0 & 0 \\ 0 & 2 & 0 \\ 0 & 0 & 0 \end{bmatrix}$ are diagonal matrices.

    More general, $A = \begin{bmatrix} a_{11} & a_{12} & \cdots & a_{1n} \\ a_{21} & a_{22} & \cdots & a_{2n} \\ \vdots & \vdots & \cdots & \vdots \\ a_{n1} & a_{n2} & \cdots & a_{nn} \end{bmatrix} = \left(a_{ij} \right)_{n \times n}$ is a diagonal

    matrix if and only if $\begin{cases} a_{ij} = 0; \text{ for } i \neq j \\ a_{ij} \neq 0; \text{ for at least one but not all } i = j \end{cases}$.

  - (b) Scalar Matrix

    A diagonal matrix in which all the diagonal elements are the same is called a *scalar matrix*, i.e. $a_{11} = a_{22} = a_{33} = \ldots = a_{nn}$.

Hence, $\begin{bmatrix} 2 & 0 \\ 0 & 2 \end{bmatrix}$ and $\begin{bmatrix} a & 0 & 0 \\ 0 & a & 0 \\ 0 & 0 & a \end{bmatrix}$ are examples of scalar

matrices.

(c) Identity or Unit Matrix

An *identity or unit matrix* is a scalar matrix in which each diagonal element is 1 (unity). An identity matrix of order n is represented by

I_n. Hence, $I_2 = \begin{bmatrix} 1 & 0 \\ 0 & 1 \end{bmatrix}$ and $I_3 = \begin{bmatrix} 1 & 0 & 0 \\ 0 & 1 & 0 \\ 0 & 0 & 1 \end{bmatrix}$ are matrices of

orders 2 and 3, respectively.

In general, $A = \begin{bmatrix} a_{11} & a_{12} & \cdots & a_{1n} \\ a_{21} & a_{22} & \cdots & a_{2n} \\ \vdots & \vdots & \cdots & \vdots \\ a_{m1} & a_{m2} & \cdots & a_{mn} \end{bmatrix} = \begin{bmatrix} a_{ij} \end{bmatrix}_{m \times n}$ is an identity

(unit) matrix if and only if $\begin{cases} a_{ij} = 0; \text{ for } i \neq j \\ a_{ij} = 1; \text{ for } i = j \end{cases}$.

4. Upper and Lower Triangular Matrix

In a square matrix $A = \begin{bmatrix} a_{ij} \end{bmatrix}_n$, if $i > j$ and $a_{ij} = 0$, then A is called an *upper triangular matrix*.

For example, $\begin{bmatrix} 1 & 5 & 9 \\ 0 & 7 & -1 \\ 0 & 0 & 5 \end{bmatrix}$ is an upper triangular matrix.

In a square matrix $A = \begin{bmatrix} a_{ij} \end{bmatrix}_n$, if $i < j$ and $a_{ij} = 0$, then A is called a *lower triangular matrix*.

For example, $\begin{bmatrix} 1 & 0 & 0 \\ 5 & 7 & 0 \\ 9 & -1 & 5 \end{bmatrix}$ is a lower triangular matrix.

5. Equal Matrices

Two matrices A and B are said to be *equal* if and only if they have the same order (i.e. m = n), and each element of matrix A is equal to the corresponding element of matrix B (i.e. for each i, j, $a_{ij} = b_{ij}$).

For example, if $A = \begin{bmatrix} 2 & 1 \\ 3 & 4 \end{bmatrix}$ and $B = \begin{bmatrix} \sqrt{4} & 1 \\ \dfrac{6}{2} & 4 \end{bmatrix}$, then A = B as the order

of matrices A and B is the same and also for each i, j, $a_{ij} = b_{ij}$.

Illustration 1.1: Find the values of a, b, c, d which satisfy the matrix equation

$$\begin{bmatrix} 2b+a & a+3 \\ c-1 & 4d-6 \end{bmatrix} = \begin{bmatrix} -7 & 0 \\ 3 & d \end{bmatrix}.$$

Solution: As per the definition of equality of matrices, comparing the elements of matrices with its corresponding terms, we get $2b+a = -7$; $a+3 = 0$; $c-1 = 3$; $4d-6 = d$.

On solving these four equations we have, $a = -3$; $b = -2$; $c = 4$; $d = 2$.

6. Negative of Matrix

The *negative of a matrix* $A = \begin{bmatrix} a_{ij} \end{bmatrix}_{m \times n}$ is the matrix formed by replacing each element in the given matrix with its additive inverse. It is represented with a negative sign with a given matrix.

For example, if $A = \begin{bmatrix} 3 & 2 & -4 \\ -1 & -2 & 5 \end{bmatrix}$ then $-A = \begin{bmatrix} -3 & -2 & 4 \\ 1 & 2 & -5 \end{bmatrix}.$

Note that for every matrix $A_{m \times n}$, the matrix $-A_{m \times n}$ holds the property that $A + (-A) = (-A) + A = 0$; where $(-A)$ is the additive inverse of A.

1.3   OPERATIONS ON MATRICES

(a) Multiplication of matrix by a scalar

kA is a matrix whose elements are the elements of matrix A, each multiplied by a constant (scalar) k.

For example, if k = 2 and $A = \begin{bmatrix} 1 & 2 & 3 \\ 1 & 2 & 1 \\ -1 & 4 & 3 \end{bmatrix}$, then $kA = \begin{bmatrix} 2 & 4 & 6 \\ 2 & 4 & 2 \\ -2 & 8 & 6 \end{bmatrix}.$

(b) Addition and Subtraction of Matrices

If two matrices A and B are of the same order m × n, then their *sum* A + B is defined as the sum of corresponding elements of A and B.

For example, $A = \begin{bmatrix} a & b \\ c & d \end{bmatrix}$ and $B = \begin{bmatrix} e & f \\ g & h \end{bmatrix}$, then $A + B = \begin{bmatrix} a+e & b+f \\ c+g & d+h \end{bmatrix}.$

The *difference* between the two elements is also carried in a similar pattern.

For example, $A = \begin{bmatrix} a & b \\ c & d \end{bmatrix}$ and $B = \begin{bmatrix} e & f \\ g & h \end{bmatrix}$, then $A - B = \begin{bmatrix} a-e & b-f \\ c-g & d-h \end{bmatrix}.$

Remark

i.   Addition of matrices of the same order holds the properties of commutativity and associativity.

ii. Addition of a matrix with the null matrix (O) also holds the property of commutativity and O has the property that $A + O = O + A = A$.

(c) Product of Matrices

Taking the product of two matrices A and B is possible if the number of columns of A is equal to the number of rows of B. Then the product AB has the same number of rows as A and the same number of columns as B. Hence, if two matrices are $A_{m \times p}$ and $B_{p \times n}$ with matrix elements a_{ij} and b_{ij}, respectively, then their product is $(AB)_{m \times n}$ with matrix elements c_{ij} determined by $c_{ij} = a_{i1}b_{1j} + a_{i2}b_{2j} + \cdots + a_{in}b_{nj}$. Hence, if $A = \begin{bmatrix} a_{11} & a_{12} \\ a_{21} & a_{22} \end{bmatrix}$ and

$B = \begin{bmatrix} b_{11} & b_{12} \\ b_{21} & b_{22} \end{bmatrix}$, then $AB = \begin{bmatrix} a_{11}b_{11} + a_{12}b_{21} & a_{11}b_{12} + a_{12}b_{22} \\ a_{21}b_{11} + a_{22}b_{21} & a_{21}b_{12} + a_{22}b_{22} \end{bmatrix}$.

Note

i.   In general, multiplication of matrices is not commutative, i.e. $AB \neq BA$.
ii. A matrix can be multiplied to itself if and only if it is a square matrix. In this case, if A is a square matrix, then its multiplication with itself gives the product $A \cdot A = A^2$. Similarly, we can define higher powers of a square matrix such as $A \cdot A^2 = A^3$, $A^2 \cdot A^2 = A^4$, etc.
iii. A is said to be pre-multiple of B, and B is said to be post-multiple of A.
iv. If a matrix A can be multiplied with an identity matrix (I), then I has the property that $AI = IA = A$.

Illustration 1.2: Find AB and BA if $A = \begin{bmatrix} -1 & 2 \\ 2 & 3 \end{bmatrix}$ and $B = \begin{bmatrix} 0 & -1 \\ 2 & -2 \end{bmatrix}$.

Solution: $AB = \begin{bmatrix} -1 & 2 \\ 2 & 3 \end{bmatrix}\begin{bmatrix} 0 & -1 \\ 2 & -2 \end{bmatrix} = \begin{bmatrix} 0+4 & 1-4 \\ 0+6 & -2-6 \end{bmatrix} = \begin{bmatrix} 4 & -3 \\ 6 & -8 \end{bmatrix}$

$BA = \begin{bmatrix} 0 & -1 \\ 2 & -2 \end{bmatrix}\begin{bmatrix} -1 & 2 \\ 2 & 3 \end{bmatrix} = \begin{bmatrix} 0-2 & 0-3 \\ -2-4 & 4-6 \end{bmatrix} = \begin{bmatrix} -2 & -3 \\ -6 & -2 \end{bmatrix}$

Illustration 1.3: Find AB if possible, given $A = \begin{bmatrix} 3 & 1 \\ 1 & 0 \\ 2 & 1 \end{bmatrix}$ and

$B = \begin{bmatrix} 1 & 2 & 3 \\ -1 & 1 & 1 \end{bmatrix}$.

Solution: AB is possible as the number of columns of $A_{3 \times 2}$ is equal to the number of rows of $B_{2 \times 3}$, and the resultant matrix AB will be of order 3×3. Hence,

$AB = \begin{bmatrix} 3 & 1 \\ 1 & 0 \\ 2 & 1 \end{bmatrix}\begin{bmatrix} 1 & 2 & 3 \\ -1 & 1 & 1 \end{bmatrix} = \begin{bmatrix} 3-1 & 6+1 & 9+1 \\ 1+0 & 2+0 & 3+0 \\ 2-1 & 4+1 & 6+1 \end{bmatrix} = \begin{bmatrix} 2 & 7 & 10 \\ 1 & 2 & 3 \\ 1 & 5 & 7 \end{bmatrix}$

Remark

Matrix multiplication holds the following properties:

i. $(AB)C = A(BC)$ (Associative Law).

ii. $\begin{aligned} A(B+C) &= AB + AC \\ (B+C)A &= BA + CA \end{aligned}$ (Distributive Law).

1.4   SOLUTION OF SYSTEM OF LINEAR EQUATIONS USING MATRICES

Let us consider the following system of equations:

$$x + y + z = 30 \tag{1.1}$$

$$-2y + z = 0 \tag{1.2}$$

$$-x + y + z = 0 \tag{1.3}$$

Now, we write the coefficients and constants of the above system of equations in a rectangular array. Note that we don't have to ignore zeroes.

$$\begin{aligned} x+y+z &= 30 \\ \text{So } -2y + z &= 0 \\ -x + y + z &= 0 \end{aligned} \Leftrightarrow \begin{bmatrix} 1 & 1 & 1 & 30 \\ 0 & -2 & 1 & 0 \\ -1 & 1 & 1 & 0 \end{bmatrix}$$

Hence, when the constants are included, the resulting matrix is referred to as an 'augmented matrix'. Augmented matrices can be used to find solutions to linear equations. Also, one can use shorthand to describe matrix operations as below:

i. R_1 represents row 1; similarly R_2 represents row 2, …
ii. Add row 1 to row 3 and replace row 3 with that sum as $R_1 + R_3 \rightarrow R_3$
iii. $R_1 \leftrightarrow R_2$ means interchanging row 1 and row 2.
iv. Multiplication of row 2 by $\dfrac{1}{2}$ is written as $\dfrac{1}{2} R_2 \rightarrow R_2$.

Step 1: Equation Form: $\begin{aligned} x + y + z &= 30 \\ -2y + z &= 0 \\ -x + y + z &= 0 \end{aligned}$

Matrix Form: $\begin{bmatrix} 1 & 1 & 1 & 30 \\ 0 & -2 & 1 & 0 \\ -1 & 1 & 1 & 0 \end{bmatrix}$

Step 2: Equation Form: Replace equation (1.3) with the sum of equations (1.1) and (1.3); we get,

$$x + y + z = 30$$

$$-2y + z = 0$$

$$2y + z = 30$$

Matrix Form: Replace row 3 with the sum of row 1 and row 3 $(R_1 + R_3 \rightarrow R_3)$; we get,

$$\begin{bmatrix} 1 & 1 & 1 & 30 \\ 0 & -2 & 1 & 0 \\ 0 & 2 & 2 & 30 \end{bmatrix}$$

Step 3: Equation Form: Multiply equation (1.2) by $\dfrac{-1}{2}$; we get,

$$x + y + z = 30$$

$$y - \frac{1}{2}z = 0$$

$$2y + 2z = 30$$

Matrix Form: Multiply row 2 by $\dfrac{-1}{2}\left(\dfrac{-1}{2}R_2 \rightarrow R_2\right)$; we get,

$$\begin{bmatrix} 1 & 1 & 1 & 30 \\ 0 & 1 & \dfrac{-1}{2} & 0 \\ 0 & 2 & 2 & 30 \end{bmatrix}$$

Step 4: Equation Form: Replace equation (1.1) with the sum of -1 times equation (1.2) and equation (1.1); we get,

$$x + \frac{3}{2}z = 30$$

$$y - \frac{1}{2}z = 0$$

$$2y + 2z = 30$$

Matrix Form: Replace row 1 with the sum of -1 times row 2 and row 1 $(-R_2 + R_1 \rightarrow R_1)$; we get,

$$\begin{bmatrix} 1 & 0 & \dfrac{3}{2} & 30 \\ 0 & 1 & \dfrac{-1}{2} & 0 \\ 0 & 2 & 2 & 0 \end{bmatrix}$$

Step 5: Equation Form: Replace equation (1.3) with the sum of $2-$ times equation (1.2) and equation (1.3); we get,

$$x + \frac{3}{2}z = 30$$

$$y - \frac{1}{2}z = 0$$

$$3z = 30$$

Matrix Form: Replace row 3 with the sum of -2 times row 2 and row 3 $(-2R_2 + R_3 \rightarrow R_3)$; we get,

$$\begin{bmatrix} 1 & 0 & \frac{3}{2} & 30 \\ 0 & 1 & \frac{-1}{2} & 0 \\ 0 & 0 & 3 & 30 \end{bmatrix}$$

Step 6: Equation Form: Multiply equation (1.3) by $\frac{1}{3}$; we get,

$$x + \frac{3}{2}z = 30$$

$$y - \frac{1}{2}z = 0$$

$$z = 10$$

Matrix Form: Multiply row 3 by $\frac{1}{3}\left(\frac{1}{3}R_3 \rightarrow R_3\right)$; we get,

$$\begin{bmatrix} 1 & 0 & \frac{3}{2} & 30 \\ 0 & 1 & \frac{-1}{2} & 0 \\ 0 & 0 & 1 & 10 \end{bmatrix}$$

Step 7: Equation Form: Replace equation (1.2) with the sum of $\left(\frac{1}{2}\right)$ times equation (1.3) and equation (1.2); we get,

$$x + \frac{3}{2}z = 30$$

$$y = 5$$

$$z = 10$$

Matrix Form: Replace row 2 with the sum of $\left(\dfrac{1}{2}\right)$ times row 3 and row 2 $\left(\dfrac{1}{2}R_3 + R_2 \to R_2\right)$; we get,

$$\begin{bmatrix} 1 & 0 & \dfrac{3}{2} & 30 \\ 0 & 1 & 0 & 5 \\ 0 & 0 & 1 & 10 \end{bmatrix}$$

Step 8: Equation Form: Replace equation (1.1) with the sum of $\left(\dfrac{-3}{2}\right)$ times equation (1.3) and equation (1.1); we get,

$$x = 15$$

$$y = 5$$

$$z = 10$$

Matrix Form: Replace row 1 with the sum of $\left(\dfrac{-3}{2}\right)$ times row 3 and row 1 $\left(\dfrac{-3}{2}R_3 + R_1 \to R_1\right)$; we get,

$$\begin{bmatrix} 1 & 0 & 0 & 15 \\ 0 & 1 & 0 & 5 \\ 0 & 0 & 1 & 10 \end{bmatrix}$$

The final matrix contains the same solution as the above in the form of equations. Notice that the first column of our matrices held the coefficients of the variable 'x', the second and third columns hold the coefficients of variables 'y' and 'z', respectively. Therefore, the first, second and third rows of the matrix can be interpreted as,

$$x + 0y + 0z = 15 \qquad\qquad\qquad x = 15$$

$$0x + y + 0z = 5 \quad \text{or more precisely} \quad y = 5$$

$$0x + 0y + z = 10 \qquad\qquad\qquad z = 10$$

1.5   ELEMENTARY ROW OPERATIONS, ROW REDUCED ECHELON FORM AND GAUSS ELIMINATION

In the earlier section, we saw how the information of a system of equations in a matrix makes sense by simply replacing the word 'equation' above with 'row'. In the context of matrices for a given system of linear equations, one can find the solution using elementary row operations. Elementary row operations give a new linear system, but the solution to the new system is the same as the older one. The following are the points of elementary row operations.

1.5.1 ELEMENTARY ROW OPERATIONS

1. Add a scalar multiple of one row to another and replace the latter row with that sum.
2. Multiplying one row by a non-zero scalar.
3. Swap the positions of the two rows.

One can use these operations as much as one wants with no change in the solution. So, the question arises when to stop. Many times, we take the original matrix, and using the elementary row operations, put it into something called *row reduced echelon form*. This form helps us to recognize whether or not the solution exists, and if it exists, what is the solution? In the previous section, when the matrices were manipulated to find solutions, unintentionally the matrix was put into a row reduced echelon form. Let's see what a row reduced echelon form is.

1.5.2 ROW REDUCED ECHELON FORM

A matrix is said to be in row reduced echelon form if it satisfies the following conditions:

1. The first non-zero entry in each non-zero row is 1.
2. If a column contains the first non-zero entry of any row, then every other entry in that column is zero.
3. The zero rows occur below all the non-zero rows.
4. Let there be r non-zero rows. If the first non-zero entry of the ith row occurs in the column k_i (I = 1, 2, 3, ..., r), then $k_1 < k_2 < ... < k_r$.

For example, $\begin{bmatrix} 1 & 0 \\ 0 & 1 \end{bmatrix}$, $\begin{bmatrix} 1 & 0 & 0 & 2 \\ 0 & 1 & 0 & 1 \\ 0 & 0 & 1 & -1 \end{bmatrix}$, $\begin{bmatrix} 0 & 0 & 1 & 0 \\ 0 & 0 & 0 & 1 \\ 0 & 0 & 0 & 0 \end{bmatrix}$, $\begin{bmatrix} 1 & 3 & 0 & 0 \\ 0 & 0 & 1 & 0 \\ 0 & 0 & 0 & 1 \end{bmatrix}$,

$\begin{bmatrix} 0 & 0 & 1 & 0 \\ 0 & 0 & 0 & 1 \\ 0 & 0 & 0 & 0 \end{bmatrix}$, $\begin{bmatrix} 1 & 0 & 3 & 0 \\ 0 & 1 & -1 & 0 \\ 0 & 0 & 0 & 1 \end{bmatrix}$, $\begin{bmatrix} 1 & 0 & 1 \\ 0 & 1 & 2 \end{bmatrix}$ are examples of row reduced

echelon form, whereas $\begin{bmatrix} 1 & 0 & 0 & 1 \\ 0 & 0 & 0 & 0 \\ 0 & 0 & 1 & 3 \end{bmatrix}$, $\begin{bmatrix} 1 & 2 & 0 & 0 \\ 0 & 0 & 3 & 0 \\ 0 & 0 & 0 & 4 \end{bmatrix}$ are examples which

are not in row reduced echelon form.

Illustration 1.4: For the following system of linear equations, put the augmented matrix into a row reduced echelon form:

$$-3x_1 - 3x_2 + 9x_3 = 12$$

$$2x_1 + 2x_2 - 4x_3 = -2$$

$$0x_1 - 2x_2 - 4x_3 = -8$$

Solution: First, convert the given linear system into an augmented matrix.

$$\begin{bmatrix} \boxed{-3} & -3 & 9 & 12 \\ 2 & 2 & -4 & -2 \\ 0 & -2 & -4 & -8 \end{bmatrix}$$

Now, we need to change the entry in a box to 1. For this, let us multiply row 1 by $\left(\dfrac{-1}{3}\right)$.

$$\begin{bmatrix} 1 & 1 & -3 & -4 \\ 2 & 2 & -4 & -2 \\ 0 & -2 & -4 & -8 \end{bmatrix} \qquad \left(\dfrac{-1}{3}R_1 \rightarrow R_1\right)$$

Now a leading 1 is created, i.e. the first entry in the first row is 1. Now using elementary row operations, our next step is to put 0 under this 1.

$$\begin{bmatrix} 1 & 1 & -3 & -4 \\ 0 & \boxed{0} & 2 & 6 \\ 0 & -2 & -4 & -8 \end{bmatrix} \qquad \left(-2R_1 + R_2 \rightarrow R_2\right)$$

Carry out elementary row operations to change the box entry from 0 to 1.

$$\begin{bmatrix} 1 & 1 & -3 & -4 \\ 0 & \boxed{-2} & -4 & -8 \\ 0 & 0 & 2 & 6 \end{bmatrix} \qquad \left(R_2 \leftrightarrow R_3\right)$$

$$\begin{bmatrix} 1 & 1 & -3 & -4 \\ 0 & 1 & 2 & 4 \\ 0 & 0 & \boxed{2} & 6 \end{bmatrix} \qquad \left(\dfrac{-1}{2}R_2 \rightarrow R_2\right)$$

Now a leading 1 has been created in the second row. Again, the box entry needs to be 1.

$$\begin{bmatrix} 1 & 1 & -3 & -4 \\ 0 & 1 & 2 & 4 \\ 0 & 0 & 1 & 3 \end{bmatrix} \qquad \left(\dfrac{1}{2}R_3 \rightarrow R_3\right)$$

This end is referred to as the forward steps. Our next task is to carry out elementary row operations and go back and put zeroes above our leading 1s. This is referred to as backward steps which are given below.

$$\begin{bmatrix} 1 & 1 & 0 & 5 \\ 0 & 1 & 0 & -2 \\ 0 & 0 & 1 & 3 \end{bmatrix} \qquad \begin{array}{l} \left(3R_3 + R_1 \rightarrow R_1\right) \\[6pt] \left(-2R_3 + R_2 \rightarrow R_2\right) \end{array}$$

$$\begin{bmatrix} 1 & 0 & 0 & 7 \\ 0 & 1 & 0 & -2 \\ 0 & 0 & 1 & 3 \end{bmatrix} \qquad \left(-R_2 + R_1 \to R_1 \right)$$

One can easily see from the above matrix that $x_1 = 7, x_2 = -2, x_3 = 3$.

Illustration 1.5: Prove that the solution of simultaneous linear equations

$$x + y + z = 6$$

$$2x - y + z = 3$$

$$x + 3y - z = 4$$

is $x = 1$, $y = 2$ and $z = 3$.

Solution: First, convert the given linear system into an augmented matrix.

$$= \begin{bmatrix} 1 & 1 & 1 & 6 \\ 2 & -1 & 1 & 3 \\ 1 & 3 & -1 & 4 \end{bmatrix}$$

Now, by $\left(R_2 + R_1 \to R_2 \right), \left(R_3 - 3R_1 \to R_3 \right)$

$$= \begin{bmatrix} 1 & 1 & 1 & 6 \\ 3 & 0 & 2 & 9 \\ -2 & 0 & -4 & -14 \end{bmatrix}$$

By $\left(\dfrac{-1}{2} R_3 \to R_3 \right)$

$$= \begin{bmatrix} 1 & 1 & 1 & 6 \\ 3 & 0 & 2 & 9 \\ 1 & 0 & 2 & 7 \end{bmatrix}$$

By $\left(R_2 - 3R_3 \to R_2 \right)$

$$= \begin{bmatrix} 1 & 1 & 1 & 6 \\ 0 & 0 & -4 & -12 \\ 1 & 0 & 2 & 7 \end{bmatrix}$$

By $\left(\dfrac{-1}{4} R_2 \to R_2 \right)$

$$= \begin{bmatrix} 1 & 1 & 1 & 6 \\ 0 & 0 & 1 & 3 \\ 1 & 0 & 2 & 7 \end{bmatrix}$$

By $\left(R_3 - 2R_2 \rightarrow R_3\right)$

$$= \begin{bmatrix} 1 & 1 & 1 & 6 \\ 0 & 0 & 1 & 3 \\ 1 & 0 & 0 & 1 \end{bmatrix}$$

By $R_1 \leftrightarrow R_3$ and $R_2 \leftrightarrow R_3$

$$= \begin{bmatrix} 1 & 0 & 0 & 1 \\ 1 & 1 & 1 & 6 \\ 0 & 0 & 1 & 3 \end{bmatrix}$$

By $R_2 - R_1 - R_3 \rightarrow R_2$

$$= \begin{bmatrix} 1 & 0 & 0 & 1 \\ 0 & 1 & 0 & 2 \\ 0 & 0 & 1 & 3 \end{bmatrix}$$

Thus, the solution is x = 1, y = 2 and z = 3.
  Hence proved.

1.5.3 GAUSS ELIMINATION

The *Gauss Elimination* method is named in honour of the great mathematician Karl Friedrich Gauss. The basic method of *Gauss Elimination* is to create leading 1s and then use elementary row operations to put zeroes above and below these leading 1s. One can do this in any order that pleases, but by following the forward and backward steps, one can make use of presence of zeroes to make the overall computation easier. This method is very efficient. It is the technique for finding the row reduced echelon form of a matrix using the above procedure, which can be abbreviated to:

1. Create a leading 1.
2. Use this leading 1 to put zeroes under it.
3. Repeat the above steps until all possible rows have leading 1s.
4. Put zeroes above these leading 1s.

Illustration 1.6: Put the given matrix into row reduced echelon form using Gauss
Elimination: $\begin{bmatrix} -2 & -4 & -2 & -10 & 0 \\ 2 & 4 & 1 & 9 & -2 \\ 3 & 6 & 1 & 13 & -4 \end{bmatrix}$.

Solution: First, we need to make the first row first column entry 1 (a leading 1).

$$\begin{bmatrix} 1 & 2 & 1 & 5 & 0 \\ 2 & 4 & 1 & 9 & -2 \\ 3 & 6 & 1 & 13 & -4 \end{bmatrix} \qquad \left(\frac{-1}{2} R_1 \rightarrow R_1 \right)$$

Now, put zeroes in the column below this newly formed leading 1.

$$\begin{bmatrix} 1 & 2 & 1 & 5 & 0 \\ 0 & 0 & \boxed{-1} & -1 & -2 \\ 0 & 0 & -2 & -2 & -4 \end{bmatrix} \qquad \begin{array}{c} \left(-2R_1 + R_2 \rightarrow R_2 \right) \\ \\ \left(-3R_1 + R_3 \rightarrow R_3 \right) \end{array}$$

The next aim is to have the leading entry 1 in the second row; i.e. we want 1 where there is −1 which is indicated in the box.

$$\begin{bmatrix} 1 & 2 & 1 & 5 & 0 \\ 0 & 0 & 1 & 1 & 2 \\ 0 & 0 & -2 & -2 & -4 \end{bmatrix} \qquad \left(-R_2 \rightarrow R_2 \right)$$

Put 0 under this leading entry 1.

$$\begin{bmatrix} 1 & 2 & 1 & 5 & 0 \\ 0 & 0 & 1 & 1 & 2 \\ 0 & 0 & 0 & 0 & 0 \end{bmatrix} \qquad \left(2R_2 + R_3 \rightarrow R_3 \right)$$

As per the procedure, we need to make the leading entry 1 in the third row, but this is not possible as one can see from the above matrix.

The next aim is to put 0 above each of the leading 1s (in this case there is only one leading 1 to deal with).

$$\begin{bmatrix} 1 & 2 & 0 & 4 & -2 \\ 0 & 0 & 1 & 1 & 2 \\ 0 & 0 & 0 & 0 & 0 \end{bmatrix} \qquad \left(-R_2 + R_1 \rightarrow R_1 \right)$$

This final matrix is in a row reduced echelon form.

MULTIPLE-CHOICE QUESTIONS

1. The order of a matrix $\begin{bmatrix} a \\ b \\ c \end{bmatrix}$ is _____.

(a) 2×1
(b) 2×2
(c) 3×1
(d) 1×3

2. The order of a matrix $\begin{bmatrix} a & b & c \end{bmatrix}$ is _____.
    (a) 1×3
    (b) 3×1
    (c) 3×3
    (d) 2×3

3. The matrix $\begin{bmatrix} 0 & 0 \\ 0 & 0 \end{bmatrix}$ is said to be _____.
    (a)   Identity
    (b)   Scalar
    (c)   Diagonal
    (d)   Null

4. Two matrices A and B are suitable for multiplication if _____.
    (a)   Number of columns in A = Number of rows in B
    (b)   Number of columns in A = Number of columns in B
    (c)   Number of rows in A = Number of rows in B
    (d)   Number of rows in A = Number of columns in B

5. If the order of matrix A is $m \times p$ and the order of matrix B is $p \times q$, then the order of matrix AB is _____.
    (a) $m \times p$
    (b) $p \times m$
    (c) $m \times q$
    (d) $q \times m$

6. In an identity matrix, all the diagonal elements are_____.
    (a)   0
    (b)   2
    (c)   1
    (d)   None of these

7. If $[a_{ij}]$ and $[b_{ij}]$ are of same order and $a_{ij} = b_{ij}$, then the matrices will be _____.
    (a)   Identity
    (b)   Null
    (c)   Unequal
    (d)   Equal

8. Matrix $\begin{bmatrix} a_{ij} \end{bmatrix}_{m \times n}$ is a row matrix if _____.
    (a) $i = 1$
    (b) $j = 1$
    (c) $m = 1$
    (d) $n = 1$

9. Matrix $\begin{bmatrix} a_{ij} \end{bmatrix}_{m \times n}$ is rectangular if _____.
    (a) $i \neq j$
    (b) $i = j$
    (c) $m = n$
    (d) $m \neq n$

10. $B = \left[b_{ij} \right]_{m \times n}$ is a scalar matrix if _____.

  (a) $b_{ij} = 0 \; \forall \, i \neq j$

  (b) $b_{ij} = k \; \forall \, i = j$

  (c) $b_{ij} = k \; \forall \, i \neq j$

  (d)   Both (a) and (b)

11. Matrix $B = \left[b_{ij} \right]_{m \times n}$ is an identity matrix if _____.

  (a) $b_{ij} = 0 \; \forall \, i = j$

  (b) $b_{ij} = 1 \; \forall \, i = j$

  (c) $b_{ij} = 0 \; \forall \, i \neq j$

  (d)   Both (b) and (c)

12. Which matrix can be a rectangular matrix from the following?

  (a)   Diagonal

  (b)   Identity

  (c)   Scalar

  (d)   None of the above

13. If $B = \left[b_{ij} \right]_{m \times n}$, then the order of kB is _____.

  (a)   m $\times$ n

  (b)   km $\times$ kn

  (c)   km $\times$ n

  (d)   m $\times$ kn

EXERCISE 1

Q.1. Write the matrix in tabular form: A = [a$_{ij}$]; where i = 1, 2, 3 and j = 1.

Q.2. Find the sums:

i. $-3 \begin{bmatrix} 4 & 2 \\ 0 & 1 \\ -5 & -1 \end{bmatrix} + 2 \begin{bmatrix} 6 & 1 \\ 0 & -3 \\ -1 & 2 \end{bmatrix}$

ii. $\begin{bmatrix} 1 & 2 & 0 \end{bmatrix} + \begin{bmatrix} 0 & -3 & 5 \end{bmatrix}$

Q.3. Find the value of X:

i. $X + \begin{bmatrix} -1 & 0 \\ 0 & 2 \end{bmatrix} = \begin{bmatrix} 2 & 6 \\ 1 & 5 \end{bmatrix} + \begin{bmatrix} -4 & -8 \\ -2 & 0 \end{bmatrix}$

ii. $X - \begin{bmatrix} 3 & -1 \\ 1 & 2 \end{bmatrix} = -2I$

Q.4. Find the products:

i. $\begin{bmatrix} 3 & -2 & 2 \end{bmatrix} \begin{bmatrix} 1 \\ 2 \\ -2 \end{bmatrix}$

ii. $\begin{bmatrix} 2 & -2 & -1 \\ 1 & 1 & -2 \\ 1 & 0 & -1 \end{bmatrix} \begin{bmatrix} -1 & -2 & 5 \\ -1 & -1 & 3 \\ -1 & -2 & 4 \end{bmatrix}$

Q.5. If $X = \begin{bmatrix} 1 & 4 \\ 2 & 1 \end{bmatrix}$, $Y = \begin{bmatrix} -3 & 2 \\ 4 & 0 \end{bmatrix}$ and $Z = \begin{bmatrix} 1 & 0 \\ 0 & 2 \end{bmatrix}$; find $X^2 + YZ$.

Q.6. If $X = \begin{bmatrix} -1 & 2 \\ 0 & 1 \end{bmatrix}$, $Y = \begin{bmatrix} 1 & 0 \\ -1 & 2 \end{bmatrix}$, then show that $(X+Y)^2 \neq X^2 + 2XY + Y^2$.

Q.7. Prove that $\begin{bmatrix} \cos\theta & 0 & -\sin\theta \\ 0 & 1 & 0 \\ \sin\theta & 0 & \cos\theta \end{bmatrix} \begin{bmatrix} \cos\theta & 0 & \sin\theta \\ 0 & 1 & 0 \\ -\sin\theta & 0 & \cos\theta \end{bmatrix} = \begin{bmatrix} 1 & 0 & 0 \\ 0 & 1 & 0 \\ 0 & 0 & 1 \end{bmatrix}$

Q.8. Show that A and B commute for $A = \begin{bmatrix} 2 & -2\sqrt{2} \\ \sqrt{2} & 2 \end{bmatrix}$ and $B = \begin{bmatrix} 2 & 2\sqrt{2} \\ -\sqrt{2} & 2 \end{bmatrix}$.

Q.9. Find x and y, if $\begin{bmatrix} 2 & 1 \\ -3 & 2 \end{bmatrix} = \begin{bmatrix} x+3 & 1 \\ -3 & 3y-4 \end{bmatrix}$.

Q.10. Find x and y, if $\begin{bmatrix} x+3 & 1 \\ -3 & 3y-4 \end{bmatrix} = \begin{bmatrix} y & 1 \\ -3 & 2x \end{bmatrix}$.

Q.11. If $A = \begin{bmatrix} 1 & 2 \\ 3 & 4 \end{bmatrix}$ and $B = \begin{bmatrix} 2 & 3 \\ 4 & 5 \end{bmatrix}$, then find:

   i. $A + B$
   ii. $A - B$
        AB

Q. 12. Find a solution to the following system of linear equations by simul-
taneously manipulating the equations and the corresponding augmented
matrices:

$$x_1 + x_2 + x_3 = 0$$

$$2x_1 + 2x_2 + x_3 = 0$$

$$-x_1 + x_2 - 2x_3 = 2$$

Q. 13. State whether the following matrices are in a row reduced echelon form
or not:

   i. $\begin{bmatrix} 1 & 1 \\ 1 & 1 \end{bmatrix}$

ii. $\begin{bmatrix} 1 & 0 & 1 \\ 0 & 1 & 2 \end{bmatrix}$

iii. $\begin{bmatrix} 1 & 1 & 1 \\ 0 & 1 & 1 \\ 0 & 0 & 1 \end{bmatrix}$

iv. $\begin{bmatrix} 1 & 0 & 0 & -5 \\ 0 & 1 & 0 & 7 \\ 0 & 0 & 1 & 3 \end{bmatrix}$

Q.14. Put the matrices in a row reduced echelon form using Gauss Elimination.

i. $\begin{bmatrix} 1 & 2 \\ -3 & -5 \end{bmatrix}$

ii. $\begin{bmatrix} -1 & 1 & 4 \\ -2 & 1 & 1 \end{bmatrix}$

iii. $\begin{bmatrix} 1 & 2 & 1 \\ 1 & 3 & 1 \\ -1 & -3 & 0 \end{bmatrix}$

iv. $\begin{bmatrix} 2 & 2 & 1 & 3 & 1 & 4 \\ 1 & 1 & 1 & 3 & 1 & 4 \end{bmatrix}$

ANSWERS TO MULTIPLE-CHOICE QUESTIONS

Answer 1: (c)
Answer 2: (a)
Answer 3: (d)
Answer 4: (a)
Answer 5: (c)
Answer 6: (c)
Answer 7: (d)
Answer 8: (c)
Answer 9: (d)
Answer 10: (d)
Answer 11: (d)
Answer 12: (d)
Answer 13: (a)

ANSWERS FOR EXERCISE 1

Answer 1: $\begin{bmatrix} a_{11} \\ a_{21} \\ a_{31} \end{bmatrix}$

Answer 2:

1. $\begin{bmatrix} 0 & 4 & 0 \\ -9 & 13 & 7 \end{bmatrix}$

ii. $\begin{bmatrix} 1 & -1 & 5 \end{bmatrix}$

Answer 3:

i. $\begin{bmatrix} -1 & -2 \\ -1 & 3 \end{bmatrix}$

1. $\begin{bmatrix} 1 & -1 \\ 1 & 0 \end{bmatrix}$

Answer 4:

i. $\begin{bmatrix} -1 \end{bmatrix}$

ii. $\begin{bmatrix} 1 & 0 & 0 \\ 0 & 1 & 0 \\ 0 & 0 & 1 \end{bmatrix}$

Answer 5: $\begin{bmatrix} 6 & 17 \\ 8 & 9 \end{bmatrix}$

Answer 9: $x = (-1), y = 2$
Answer 10: $x = (-5), y = 2$
Answer 11:

i. $A + B = \begin{bmatrix} 3 & 5 \\ 7 & 9 \end{bmatrix}$

ii. $A - B = \begin{bmatrix} -1 & -1 \\ -1 & -1 \end{bmatrix}$

iii. $AB = \begin{bmatrix} 10 & 13 \\ 22 & 29 \end{bmatrix}$

Answer 12:

$x_1 = -1$

$x_2 = 1$ and in the form of matrix $\begin{bmatrix} 1 & 0 & 0 & -1 \\ 0 & 1 & 0 & 1 \\ 0 & 0 & 1 & 0 \end{bmatrix}$

$x_3 = 0$

Answer 13:

i. No
ii. Yes
iii. No
iv. Yes

Answer 14:

i. $\begin{bmatrix} 1 & 0 \\ 0 & 1 \end{bmatrix}$

ii. $\begin{bmatrix} 1 & 0 & 3 \\ 0 & 1 & 7 \end{bmatrix}$

iii. $\begin{bmatrix} 1 & 0 & 0 \\ 0 & 1 & 0 \\ 0 & 0 & 1 \end{bmatrix}$

iv. $\begin{bmatrix} 1 & 1 & 0 & 0 & 0 & 0 \\ 0 & 0 & 1 & 3 & 1 & 4 \end{bmatrix}$

2 Determinants

2.1 INTRODUCTION

The determinant of a matrix, which is the characteristic of a matrix, is a number (a scalar quantity) obtained from the elements of a matrix by a specified operation. The determinant can be calculated or defined only for a square matrix. It is represented as det A or $|A|$ for a given square matrix A.

The determinant of an order 2×2 matrix, $A = \begin{bmatrix} a_{11} & a_{12} \\ a_{21} & a_{22} \end{bmatrix}$, is given by

$$\det A = |A| = \begin{vmatrix} a_{11} & a_{12} \\ a_{21} & a_{22} \end{vmatrix} = a_{11}a_{22} - a_{12}a_{21}$$

The determinant of an order 3×3 matrix, $A = \begin{bmatrix} a_{11} & a_{12} & a_{13} \\ a_{21} & a_{22} & a_{23} \\ a_{31} & a_{32} & a_{33} \end{bmatrix}$, is given by

$$\det A = |A| = \begin{vmatrix} a_{11} & a_{12} & a_{13} \\ a_{21} & a_{22} & a_{23} \\ a_{31} & a_{32} & a_{33} \end{vmatrix}$$

$$= a_{11} \begin{vmatrix} a_{22} & a_{23} \\ a_{32} & a_{33} \end{vmatrix} - a_{12} \begin{vmatrix} a_{21} & a_{23} \\ a_{31} & a_{33} \end{vmatrix} + a_{13} \begin{vmatrix} a_{21} & a_{22} \\ a_{31} & a_{32} \end{vmatrix}$$

$$= a_{11}(a_{22}a_{33} - a_{23}a_{32}) - a_{12}(a_{21}a_{33} - a_{23}a_{31}) + a_{13}(a_{21}a_{32} - a_{22}a_{31})$$

Note: Each small determinant in the sum on the RHS is the determinant of a submatrix of A, obtained by deleting the associated row and column of A. These small determinants are called *minors*. The sign '+' or '−' is based on $(-1)^{i+j} a_{ij}$, where i and j represent row and column, respectively.

Illustration 2.1: Find the determinant $|A|$ if $A = \begin{bmatrix} 2 & 1 \\ -1 & 3 \end{bmatrix}$.

Solution: $|A| = 6 - (-1) = 7$.

Illustration 2.2: Find the determinant $|A|$ if $A = \begin{bmatrix} 1 & 2 & 3 \\ 0 & 1 & -2 \\ 3 & 2 & 4 \end{bmatrix}$.

Solution: $|A| = 1(4+4) - 2(0+6) + 3(0-3) = 8 - 12 - 9 = -13$.

2.2　MINOR AND COFACTOR OF ELEMENT

In a given determinant |A|, the *minor* M_{ij} of the element a_{ij} is the determinant of order $(n-1 \times n-1)$ obtained by deleting the ith row and j^{th} column of $A_{n \times n}$.

For example, in a given determinant

$$|A| = \begin{vmatrix} a_{11} & a_{12} & a_{13} \\ a_{21} & a_{22} & a_{23} \\ a_{31} & a_{32} & a_{33} \end{vmatrix} \qquad (2.1)$$

The minor of the element a_{11} is $M_{11} = \begin{vmatrix} a_{22} & a_{23} \\ a_{32} & a_{33} \end{vmatrix}$.

The minor of the element a_{12} is $M_{12} = \begin{vmatrix} a_{21} & a_{23} \\ a_{31} & a_{33} \end{vmatrix}$.

The minor of the element a_{13} is $M_{13} = \begin{vmatrix} a_{21} & a_{22} \\ a_{31} & a_{32} \end{vmatrix}$, and similarly it can be calcu-

lated for the other elements also.

Now, the *cofactor* of the element a_{ij} of the given matrix A is the scalar quantity $C_{ij} = (-1)^{i+j} M_{ij}$.

For example, for a given matrix $A = \begin{bmatrix} a_{11} & a_{12} & a_{13} \\ a_{21} & a_{22} & a_{23} \\ a_{31} & a_{32} & a_{33} \end{bmatrix}$

The cofactor of the element a_{11} is $C_{11} = (-1)^{1+1} M_{11} = M_{11} = \begin{vmatrix} a_{22} & a_{23} \\ a_{32} & a_{33} \end{vmatrix}$.

The cofactor of the element a_{12} is $C_{12} = (-1)^{1+2} M_{12} = -M_{12} = -\begin{vmatrix} a_{21} & a_{23} \\ a_{31} & a_{33} \end{vmatrix}$.

The cofactor of the element a_{13} is $C_{13} = (-1)^{1+3} M_{13} = M_{13} = \begin{vmatrix} a_{21} & a_{22} \\ a_{31} & a_{32} \end{vmatrix}$, and simi-

larly it can be calculated for the other elements.

Note: Using minors or cofactors, the value of the determinant in equation (2.1) can also be calculated as $a_{11}M_{11} - a_{12}M_{12} + a_{13}M_{13}$ or $a_{11}C_{11} + a_{12}C_{12} + a_{13}C_{13}$, respectively. Thus, det A is the sum of any row or column multiplied by their corresponding cofactors. Also, the value of the determinant can be found by expanding it from any of the row or column.

Illustration 2.3: Find the minor and cofactor of element 4 for the matrix $\begin{bmatrix} 1 & 2 & 3 \\ 4 & 5 & 6 \\ 7 & 8 & 9 \end{bmatrix}$.

Solution: Minor of $4 = \begin{vmatrix} 2 & 3 \\ 8 & 9 \end{vmatrix} = 18 - 24 = (-6)$

Cofactor of $4 = (-1)^{2+1} M_{21} = (-1)^3 (\text{Minor of } 4) = -(-6) = 6$

Illustration 2.4: Find the value of $\begin{vmatrix} 3 & 4 & -1 \\ 2 & 0 & 7 \\ 1 & -3 & -2 \end{vmatrix}$ by expanding the first column.

$$\begin{vmatrix} 3 & 4 & -1 \\ 2 & 0 & 7 \\ 1 & -3 & -2 \end{vmatrix} = 3 \begin{vmatrix} 0 & 7 \\ -3 & -2 \end{vmatrix} - 2 \begin{vmatrix} 4 & -1 \\ -3 & -2 \end{vmatrix} + 1 \begin{vmatrix} 4 & -1 \\ 0 & 7 \end{vmatrix}$$

$$= 3(0 + 21) - 2(-8 - 3) + 1(28 + 0)$$

Solution:
$$= 3(21) - 2(-11) + 1(28)$$

$$= 63 + 22 + 28$$

$$= 113$$

2.3   PROPERTIES OF DETERMINANTS

The following are the properties which are most useful in evaluating determinants:

1. On interchanging the corresponding rows and columns, the value of the determinant does not change.

    For example, consider a determinant

$$|A| = \begin{vmatrix} a_1 & b_1 & c_1 \\ a_2 & b_2 & c_2 \\ a_3 & b_3 & c_3 \end{vmatrix}$$

$$= a_1 (b_2 c_3 - c_2 b_3) - b_1 (a_2 c_3 - c_2 a_3) + c_1 (a_2 b_3 - b_2 a_3) \qquad (2.2)$$

Now, considering the determinant by interchanging the corresponding row with the corresponding column of determinant A, we have,

$$|B| = \begin{vmatrix} a_1 & a_2 & a_3 \\ b_1 & b_2 & b_3 \\ c_1 & c_2 & c_3 \end{vmatrix}$$

Expanding over the first column, we get,

$$= a_1 (b_2 c_3 - c_2 b_3) - b_1 (a_2 c_3 - c_2 a_3) + c_1 (a_2 b_3 - b_2 a_3) \qquad (2.3)$$

So from equations (2.2) and (2.3), $|B| = |A|$

2. The sign of the determinant changes if two rows or two columns of a determinant are interchanged.

For example, consider a determinant

$$|A| = \begin{vmatrix} a_1 & b_1 & c_1 \\ a_2 & b_2 & c_2 \\ a_3 & b_3 & c_3 \end{vmatrix}$$

$$= a_1 (b_2 c_3 - c_2 b_3) - b_1 (a_2 c_3 - c_2 a_3) + c_1 (a_2 b_3 - b_2 a_3)$$

Now, considering the determinant by interchanging two rows of determinant A we get, $|B| = \begin{vmatrix} a_2 & b_2 & c_2 \\ a_1 & b_1 & c_1 \\ a_3 & b_3 & c_3 \end{vmatrix}$

Expanding over the second row, we get,

$$= -a_1 (b_2 c_3 - c_2 b_3) + b_1 (a_2 c_3 - c_2 a_3) - c_1 (a_2 b_3 - b_2 a_3)$$

$$= -\left(a_1 (b_2 c_3 - c_2 b_3) - b_1 (a_2 c_3 - c_2 a_3) + c_1 (a_2 b_3 - b_2 a_3) \right)$$

$$= - \begin{vmatrix} a_1 & b_1 & c_1 \\ a_2 & b_2 & c_2 \\ a_3 & b_3 & c_3 \end{vmatrix}$$

$$= -|A|$$

3. The value of a determinant is zero if every element of a row or a column of a determinant is zero.

For example, consider a determinant

$$|A| = \begin{vmatrix} 0 & 0 & 0 \\ a_2 & b_2 & c_2 \\ a_3 & b_3 & c_3 \end{vmatrix}$$

$$= 0 (b_2 c_3 - c_2 b_3) - 0 (a_2 c_3 - c_2 a_3) + 0 (a_2 b_3 - b_2 a_3)$$

$$= 0$$

4. The value of a determinant is zero if two rows or columns of a determinant are identical.

For example, consider a determinant

$$A = \begin{vmatrix} a_1 & b_1 & c_1 \\ a_1 & b_1 & c_1 \\ a_3 & b_3 & c_3 \end{vmatrix}$$

$$= a_1 \left(b_1 c_3 - c_1 b_3 \right) - b_1 \left(a_1 c_3 - c_1 a_3 \right) + c_1 \left(a_1 b_3 - b_1 a_3 \right)$$

$$= a_1 b_1 c_3 - a_1 c_1 b_3 - a_1 b_1 c_3 + b_1 c_1 a_3 + a_1 c_1 b_3 - b_1 c_1 a_3$$

$$= 0$$

5. If each element of a row or column of a determinant is multiplied by the same constant k, the value of the determinant is also multiplied by that same constant k.

For example, consider a determinant

$$A = \begin{vmatrix} a_1 & b_1 & c_1 \\ a_2 & b_2 & c_2 \\ a_3 & b_3 & c_3 \end{vmatrix}$$

Now consider $B = \begin{vmatrix} ka_1 & kb_1 & kc_1 \\ a_2 & b_2 & c_2 \\ a_3 & b_3 & c_3 \end{vmatrix}$

$$= ka_1 \left(b_2 c_3 - c_2 b_3 \right) - kb_1 \left(a_2 c_3 - c_2 a_3 \right) + kc_1 \left(a_2 b_3 - b_2 a_3 \right)$$

$$= k \left(a_1 \left(b_2 c_3 - c_2 b_3 \right) - b_1 \left(a_2 c_3 - c_2 a_3 \right) + c_1 \left(a_2 b_3 - b_2 a_3 \right) \right)$$

$$= k |A|$$

6. If each element of any row or column is added to (or subtracted from) a constant multiple of the corresponding element of any other row or column, then the value of a determinant is not changed.

For example, consider a determinant

$$A = \begin{vmatrix} a_1 & b_1 & c_1 \\ a_2 & b_2 & c_2 \\ a_3 & b_3 & c_3 \end{vmatrix}$$

Now, consider a determinant, $B = \begin{vmatrix} a_1 + ka_2 & b_1 + kb_2 & c_1 + kc_2 \\ a_2 & b_2 & c_2 \\ a_3 & b_3 & c_3 \end{vmatrix}$

$$= \left(a_1 + ka_2\right)\left(b_2c_3 - c_2b_3\right) - \left(b_1 + kb_2\right)\left(a_2c_3 - c_2a_3\right) + \left(c_1 + kc_1\right)\left(a_2b_3 - b_2a_3\right)$$

$$= \left[a_1\left(b_2c_3 - c_2b_3\right) - b_1\left(a_2c_3 - c_2a_3\right) + c_1\left(a_2b_3 - b_2a_3\right)\right]$$

$$+ \left[ka_2\left(b_2c_3 - c_2b_3\right) - kb_2\left(a_2c_3 - c_2a_3\right) + kc_2\left(a_2b_3 - b_2a_3\right)\right]$$

$$= \begin{vmatrix} a_1 & b_1 & c_1 \\ a_2 & b_2 & c_2 \\ a_3 & b_3 & c_3 \end{vmatrix} + k \begin{vmatrix} a_2 & b_2 & c_2 \\ a_2 & b_2 & c_2 \\ a_3 & b_3 & c_3 \end{vmatrix}$$

$$= |A| + k(0) \left[\text{Since, two rows are identical so value is zero}\right]$$

$$= |A|$$

7. The determinant of a diagonal matrix is equal to the product of its diagonal elements.

   For example, consider a determinant

$$|A| = \begin{vmatrix} 2 & 0 & 0 \\ 0 & -7 & 0 \\ 0 & 0 & 4 \end{vmatrix}$$

$$= 2(-28 - 0) - 0 + 0$$

$$= -56 \text{ (which is also the product of diagonal elements)}.$$

8. The determinant of the product of two matrices is equal to (the same as) that of the product of the determinant of two matrices.

   Here, mathematically we need to show that $|AB| = |A||B|$.

   For this, let us consider $A = \begin{vmatrix} a_{11} & a_{12} \\ a_{21} & a_{22} \end{vmatrix}$ and $B = \begin{vmatrix} b_{11} & b_{12} \\ b_{21} & b_{22} \end{vmatrix}$.

   Then, $AB = \begin{vmatrix} a_{11}b_{11} + a_{12}b_{21} & a_{11}b_{12} + a_{12}b_{22} \\ a_{21}b_{11} + a_{22}b_{21} & a_{21}b_{12} + a_{22}b_{22} \end{vmatrix}$

$$|AB| = \left[\left(a_{11}b_{11} + a_{12}b_{21}\right)\left(a_{21}b_{12} + a_{22}b_{22}\right) - \left(a_{11}b_{12} + a_{12}b_{22}\right)\left(a_{21}b_{11} + a_{22}b_{21}\right)\right]$$

On simplification, we get,

$$|AB| = a_{11}b_{11}a_{22}b_{22} + a_{12}b_{21}a_{21}b_{12} - a_{11}b_{12}a_{22}b_{21} - a_{12}b_{22}a_{21}b_{11} \qquad (2.4)$$

Also, $|A| = a_{11}a_{22} - a_{12}a_{21}$ and $|B| = b_{11}b_{22} - b_{12}b_{21}$

Therefore, $|A||B| = \left[\left(a_{11}a_{22} - a_{12}a_{21}\right)\left(b_{11}b_{22} - b_{12}b_{21}\right)\right]$

$$|A||B| = a_{11}b_{11}a_{22}b_{22} + a_{12}b_{21}a_{21}b_{12} - a_{11}b_{12}a_{22}b_{21} - a_{12}b_{22}a_{21}b_{11} \qquad (2.5)$$

Hence, from equations (2.4) and (2.5) we have obtained,

$$|AB| = |A||B|$$

9. The determinant can be expressed as the sum of two other determinants if the determinant in each element in any row or column consists of two terms.

For this, let us consider the determinant $\begin{vmatrix} a_1 + \alpha_1 & b_1 & c_1 \\ a_2 + \alpha_2 & b_2 & c_2 \\ a_3 + \alpha_3 & b_3 & c_3 \end{vmatrix}$.

Here, we need to show that

$$\begin{vmatrix} a_1 + \alpha_1 & b_1 & c_1 \\ a_2 + \alpha_2 & b_2 & c_2 \\ a_3 + \alpha_3 & b_3 & c_3 \end{vmatrix} = \begin{vmatrix} a_1 & b_1 & c_1 \\ a_2 & b_2 & c_2 \\ a_3 & b_3 & c_3 \end{vmatrix} + \begin{vmatrix} \alpha_1 & b_1 & c_1 \\ \alpha_2 & b_2 & c_2 \\ \alpha_3 & b_3 & c_3 \end{vmatrix}.$$

To prove this, we will expand the LHS by the first column,

$$\text{LHS} = (a_1 + \alpha_1)(b_2c_3 - c_2b_3) - (a_2 + \alpha_2)(b_1c_3 - c_1b_3) + (a_3 + \alpha_3)(b_1c_2 - c_1b_2)$$

$$= \left[a_1(b_2c_3 - c_2b_3) - a_2(b_1c_3 - c_1b_3) + a_3(b_1c_2 - c_1b_2) \right]$$

$$+ \left[\alpha_1(b_2c_3 - c_2b_3) - \alpha_2(b_1c_3 - c_1b_3) + \alpha_3(b_1c_2 - c_1b_2) \right]$$

$$= \begin{vmatrix} a_1 & b_1 & c_1 \\ a_2 & b_2 & c_2 \\ a_3 & b_3 & c_3 \end{vmatrix} + \begin{vmatrix} \alpha_1 & b_1 & c_1 \\ \alpha_2 & b_2 & c_2 \\ \alpha_3 & b_3 & c_3 \end{vmatrix}$$

$$= \text{RHS}$$

Hence proved.

Similarly, one can also prove the following results,

i.
$$\begin{vmatrix} a_1 + \alpha_1 & b_1 + \beta_1 & c_1 \\ a_2 + \alpha_2 & b_2 + \beta_2 & c_2 \\ a_3 + \alpha_3 & b_3 + \beta_3 & c_3 \end{vmatrix} = \begin{vmatrix} a_1 & b_1 & c_1 \\ a_2 & b_2 & c_2 \\ a_3 & b_3 & c_3 \end{vmatrix} + \begin{vmatrix} \alpha_1 & b_1 & c_1 \\ \alpha_2 & b_2 & c_2 \\ \alpha_3 & b_3 & c_3 \end{vmatrix}$$

$$+ \begin{vmatrix} a_1 & \beta_1 & c_1 \\ a_2 & \beta_2 & c_2 \\ a_3 & \beta_3 & c_3 \end{vmatrix} + \begin{vmatrix} \alpha_1 & \beta_1 & c_1 \\ \alpha_2 & \beta_2 & c_2 \\ \alpha_3 & \beta_3 & c_3 \end{vmatrix}$$

ii. $\begin{vmatrix} a_1+\alpha_1 & b_1+\beta_1 & c_1+\gamma_1 \\ a_2+\alpha_2 & b_2+\beta_2 & c_2+\gamma_2 \\ a_3+\alpha_3 & b_3+\beta_3 & c_3+\gamma_3 \end{vmatrix} = \begin{vmatrix} a_1 & b_1 & c_1 \\ a_2 & b_2 & c_2 \\ a_3 & b_3 & c_3 \end{vmatrix} + $ sum of six determinants $ + \begin{vmatrix} \alpha_1 & \beta_1 & \gamma_1 \\ \alpha_2 & \beta_2 & \gamma_2 \\ \alpha_3 & \beta_3 & \gamma_3 \end{vmatrix}$

Illustration 2.5: Show that $\begin{vmatrix} 6 & 3 & 2 \\ -4 & 2 & 8 \\ -2 & 1 & 4 \end{vmatrix} = 0$ without expansion.

Solution: $2\begin{vmatrix} 6 & 3 & 2 \\ -2 & 1 & 4 \\ -2 & 1 & 4 \end{vmatrix}$ (since 2 is taken common from the second row of the

given determinant)

= 0 (as two rows of the above determinant are identical)

Illustration 2.6: Prove that $\begin{vmatrix} 1 & a & bc \\ 1 & b & ca \\ 1 & c & ab \end{vmatrix} = \begin{vmatrix} 1 & a & a^2 \\ 1 & b & b^2 \\ 1 & c & c^2 \end{vmatrix}$

Solution: LHS $= \begin{vmatrix} 1 & a & bc \\ 1 & b & ca \\ 1 & c & ab \end{vmatrix}$

$= \dfrac{1}{abc}\begin{vmatrix} a & a^2 & abc \\ b & b^2 & abc \\ c & c^2 & abc \end{vmatrix}$ (multiplying the first row by a, second row by b, third row

by c)

$= \dfrac{abc}{abc}\begin{vmatrix} a & a^2 & 1 \\ b & b^2 & 1 \\ c & c^2 & 1 \end{vmatrix}$ (taking abc common from the third column)

$= -\begin{vmatrix} 1 & a^2 & a \\ 1 & b^2 & b \\ 1 & c^2 & c \end{vmatrix}$ (interchanging columns 1 and 3)

$= \begin{vmatrix} 1 & a & a^2 \\ 1 & b & b^2 \\ 1 & c & c^2 \end{vmatrix}$ (interchanging columns 2 and 3)

= RHS
Hence proved.

Illustration 2.7: Verify that $\begin{vmatrix} 1 & a & a^2 \\ 1 & b & b^2 \\ 1 & c & c^2 \end{vmatrix} = (a-b)(b-c)(c-a).$

Solution: LHS $= \begin{vmatrix} 1 & a & a^2 \\ 1 & b & b^2 \\ 1 & c & c^2 \end{vmatrix}$

$= \begin{vmatrix} 0 & a-b & a^2-b^2 \\ 0 & b-c & b^2-c^2 \\ 1 & c & c^2 \end{vmatrix}$ (first row minus second row and second row minus third

row)

$= \begin{vmatrix} 0 & a-b & (a-b)(a+b) \\ 0 & b-c & (b-c)(b+c) \\ 1 & c & c^2 \end{vmatrix}$

$= (a-b)(b-c) \begin{vmatrix} 0 & 1 & a+b \\ 0 & 1 & b+c \\ 1 & c & c^2 \end{vmatrix}$

(since a − b and b − c are taken as common from the first and second row in the above determinant)

$= (a-b)(b-c)(b+c-a-b)$ (expanding by the first column)

$= (a-b)(b-c)(c-a)$

= RHS
Hence proved.

2.4   SOLUTION OF LINEAR EQUATIONS BY DETERMINANTS (CRAMER'S RULE)

2.4.1   SOLUTION FOR A SYSTEM OF LINEAR EQUATIONS IN TWO VARIABLES

Let us consider a system of linear equations in two variables x and y,

$$a_1x + b_1y = c_1 \tag{2.6}$$

$$a_2x + b_2y = c_2 \tag{2.7}$$

Multiplying equation (2.6) by b_2 and equation (2.7) by b_1 and subtracting, we get,

$$(a_1b_2 - a_2b_1)x + (b_1b_2 - b_2b_1)y = b_2c_1 - b_1c_2$$

$$\therefore \left(a_1 b_2 - a_2 b_1\right) x = b_2 c_1 - b_1 c_2$$

Note: '$\therefore$' indicates 'therefore'.

$$\therefore \; x = \frac{b_2 c_1 - b_1 c_2}{a_1 b_2 - a_2 b_1} \tag{2.8}$$

Again, multiplying equation (2.6) by a_2 and equation (2.7) by a_1 and subtracting, we get,

$$\left(a_1 a_2 - a_2 a_1\right) x + \left(a_2 b_1 - a_1 b_2\right) y = a_2 c_1 - a_1 c_2$$

$$\therefore \; \left(a_2 b_1 - a_1 b_2\right) y = a_2 c_1 - a_1 c_2$$

$$\therefore \; y = \frac{a_2 c_1 - a_1 c_2}{a_2 b_1 - a_1 b_2}$$

$$\therefore \; y = \frac{a_1 c_2 - a_2 c_1}{a_1 b_2 - a_2 b_1} \tag{2.9}$$

From equations (2.8) and (2.9), it is observed that both x and y have the same denominators. Hence, the system of equations (2.5) and (2.6) has a solution only when $a_1 b_2 - a_2 b_1 \neq 0$.

The solutions x and y of the system of equations can also be written in the form of determinant as $x = \dfrac{\begin{vmatrix} c_1 & b_1 \\ c_2 & b_2 \end{vmatrix}}{\begin{vmatrix} a_1 & b_1 \\ a_2 & b_2 \end{vmatrix}}$ and $y = \dfrac{\begin{vmatrix} a_1 & c_1 \\ a_2 & c_2 \end{vmatrix}}{\begin{vmatrix} a_1 & b_1 \\ a_2 & b_2 \end{vmatrix}}.$

This result is known as *Cramer's Rule*.

Here, $\begin{vmatrix} a_1 & b_1 \\ a_2 & b_2 \end{vmatrix} = |A|$ is the determinant of the coefficient of x and y in the given equations (2.5) and (2.6).

If $\begin{vmatrix} c_1 & b_1 \\ c_2 & b_2 \end{vmatrix} = |A_x|$ and $\begin{vmatrix} a_1 & c_1 \\ a_2 & c_2 \end{vmatrix} = |A_y|$, then $x = \dfrac{|A_x|}{|A|}$ and $y = \dfrac{|A_y|}{|A|}.$

2.4.2   Solution for a System of Linear Equations in Three Variables

Let us consider a system of linear equations in three variables x, y and z,

$$a_1 x + b_1 y + c_1 z = d_1$$

$$a_2 x + b_2 y + c_2 z = d_2$$

$$a_3x + b_3y + c_3z = d_3$$

Thus, the determinant of coefficients is $|A| = \begin{vmatrix} a_1 & b_1 & c_1 \\ a_2 & b_2 & c_2 \\ a_3 & b_3 & c_3 \end{vmatrix}$ if $|A| \neq 0$.

Then by *Cramer's Rule*, the value of variables x, y and z is,

$$x = \frac{\begin{vmatrix} d_1 & b_1 & c_1 \\ d_2 & b_2 & c_2 \\ d_3 & b_3 & c_3 \end{vmatrix}}{|A|} = \frac{|A_x|}{|A|},$$

$$y = \frac{\begin{vmatrix} a_1 & d_1 & c_1 \\ a_2 & d_2 & c_2 \\ a_3 & d_3 & c_3 \end{vmatrix}}{|A|} = \frac{|A_y|}{|A|} \text{ and}$$

$$z = \frac{\begin{vmatrix} a_1 & b_1 & d_1 \\ a_2 & b_2 & d_2 \\ a_3 & b_3 & d_3 \end{vmatrix}}{|A|} = \frac{|A_z|}{|A|}$$

Illustration 2.8: Solve the following system using Cramer's Rule:

$$x - y = 2$$

$$x + 4y = 5$$

Solution: By Cramer's Rule,

$$|A_x| = \begin{vmatrix} 2 & -1 \\ 5 & 4 \end{vmatrix} = 8 + 5 = 13$$

$$|A_y| = \begin{vmatrix} 1 & 2 \\ 1 & 5 \end{vmatrix} = 5 - 2 = 3$$

$$|A| = \begin{vmatrix} 1 & -1 \\ 1 & 4 \end{vmatrix} = 4 + 1 = 5$$

Hence, $x = \frac{|A_x|}{|A|} = \frac{13}{5}$ and $y = \frac{|A_y|}{|A|} = \frac{3}{5}$

So the solution set is $\left\{\left(\dfrac{13}{5},\dfrac{3}{5}\right)\right\}$.

Illustration 2.9: Solve the following system using Cramer's Rule:

$$x + y + z = 9$$

$$2x + 5y + 7z = 52$$

$$2x + y - z = 0$$

Solution: By Cramer's Rule,

$$|A_x| = \begin{vmatrix} 9 & 1 & 1 \\ 52 & 5 & 7 \\ 0 & 1 & -1 \end{vmatrix} = 9(-12) - (-52) + 52 = -108 + 104 = -4$$

$$|A_y| = \begin{vmatrix} 1 & 9 & 1 \\ 2 & 52 & 7 \\ 2 & 0 & -1 \end{vmatrix} = -52 - 9(-16) + (-104) = -52 + 144 - 104 = -12$$

$$|A_z| = \begin{vmatrix} 1 & 1 & 9 \\ 2 & 5 & 52 \\ 2 & 1 & 0 \end{vmatrix} = -52 - (-104) + 9(-8) = -52 + 104 - 72 = -20$$

$$|A| = \begin{vmatrix} 1 & 1 & 1 \\ 2 & 5 & 7 \\ 2 & 1 & -1 \end{vmatrix} = -12 - (-16) + (-8) = -12 + 16 - 8 = -4$$

Hence, $x = \dfrac{|A_x|}{|A|} = \dfrac{-4}{-4} = 1$, $y = \dfrac{|A_y|}{|A|} = \dfrac{-12}{-4} = 3$ and $z = \dfrac{|A_z|}{|A|} = \dfrac{-20}{-4} = 5$

So the solution set is $\{(1,3,5)\}$.

MULTIPLE-CHOICE QUESTIONS

1. If two rows of a determinant are identical, then its value is _____.
   (a) 1
   (b) 0
   (c) −1
   (d) None of these

2. The cofactor of 4 in the matrix $\begin{bmatrix} 2 & 3 & 4 \\ 0 & 1 & -1 \\ 2 & 0 & 1 \end{bmatrix}$ is _____.

(a)   −2
(b)   2
(c)   3
(d)   4

3. If all the elements of a row or a column are zero, then the value of the determinant is _____.

(a)   1
(b)   2
(c)   0
(d)   None of these

4. The determinant of a diagonal matrix is equal to the _____ of its diagonal elements.

(a)   Sum
(b)   Difference
(c)   Product
(d)   None of these

5. The value of the determinant $\begin{vmatrix} 1 & 0 & 0 \\ 0 & 2 & 0 \\ 0 & 0 & 4 \end{vmatrix}$ is _____.

(a)   1
(b)   2
(c)   4
(d)   8

6. The value of the determinant $\begin{vmatrix} 1 & 2 \\ 3 & 4 \end{vmatrix}$ is _____.

(a)   −2
(b)   2
(c)   1
(d)   0

EXERCISE 2

Q.1. Expand the determinants:

i. $\begin{vmatrix} -10 & 0 & 0 \\ 0 & 10 & 0 \\ 0 & 0 & -10 \end{vmatrix}$

ii. $\begin{vmatrix} x & y & 1 \\ x & y & 1 \\ 1 & 1 & 1 \end{vmatrix}$

iii. $\begin{vmatrix} 1 & 2 & 0 \\ 3 & -1 & 4 \\ -2 & 1 & 3 \end{vmatrix}$

iv. $\begin{vmatrix} 1 & 2 & -2 \\ -1 & 1 & -3 \\ 2 & 4 & -1 \end{vmatrix}$

Q.2. Verify the following without expansion:

i. $\begin{vmatrix} 1 & 2 & 1 \\ 0 & 2 & 3 \\ 2 & -1 & 2 \end{vmatrix} = \begin{vmatrix} 1 & 2 & 1 \\ 0 & 2 & 3 \\ 0 & -5 & 0 \end{vmatrix}$

ii. $\begin{vmatrix} -2 & 1 & 0 \\ 3 & 5 & 4 \\ -8 & 4 & 0 \end{vmatrix} = 0$

iii. $\begin{vmatrix} a-b & b-c & c-a \\ b-c & c-a & a-b \\ c-a & a-b & b-c \end{vmatrix} = 0$

iv. $\begin{vmatrix} x+1 & x+2 & x+3 \\ x+4 & x+5 & x+6 \\ x+7 & x+8 & x+9 \end{vmatrix} = 0$

v. $\begin{vmatrix} bc & ca & ab \\ a^3 & b^3 & c^3 \\ \frac{1}{a} & \frac{1}{b} & \frac{1}{c} \end{vmatrix} = 0$

vi. $\begin{vmatrix} a & b & c \\ d & e & f \\ g & h & k \end{vmatrix} = \begin{vmatrix} e & b & h \\ d & a & g \\ f & c & k \end{vmatrix}$

vii. $\begin{vmatrix} 1 & 2 & 3 \\ 4 & 5 & 6 \\ 7 & 8 & 9 \end{vmatrix} = 0$

Q.3. Prove that

$$\begin{vmatrix} a_1 & a_2 & a_3 \\ b_1 & b_2 & b_3 \\ c_1x+d_1 & c_2x+d_2 & c_3x+d_3 \end{vmatrix} = x\begin{vmatrix} a_1 & a_2 & a_3 \\ b_1 & b_2 & b_3 \\ c_1 & c_2 & c_3 \end{vmatrix} + \begin{vmatrix} a_1 & a_2 & a_3 \\ b_1 & b_2 & b_3 \\ d_1 & d_2 & d_3 \end{vmatrix}.$$

Q.4. Show that:

i. $\begin{vmatrix} x & a & a \\ a & x & a \\ a & a & x \end{vmatrix} = (x-a)^2(2a+x)$

ii. $\begin{vmatrix} a+x & a & a \\ a & a+x & a \\ a & a & a+x \end{vmatrix} = x^2(3a+x)$

iii. $\begin{vmatrix} 0 & a & b \\ -a & 0 & c \\ -b & -c & 0 \end{vmatrix} = 0$

iv. $\begin{vmatrix} a & b & c \\ a & a+b & a+b+c \\ a & 2a+b & 3a+2b+c \end{vmatrix} = a^3$

v. $\begin{vmatrix} a-b-c & 2a & 2a \\ 2b & b-c-a & 2b \\ 2c & 2c & c-a-b \end{vmatrix} = (a+b+c)^3$

vi. $\begin{vmatrix} 1 & 1 & 1 \\ bc & ca & ab \\ b+c & c+a & a+b \end{vmatrix} = (a-b)(b-c)(c-a)$

vii. $\begin{vmatrix} a+\lambda & b & c \\ a & b+\lambda & c \\ a & b & c+\lambda \end{vmatrix} = \lambda^2(a+b+c+\lambda)$

Q.5. Find the values of x if:

i. $\begin{vmatrix} 1 & 2 & 1 \\ 2 & x & 2 \\ 3 & 6 & x \end{vmatrix} = 0$

ii. $\begin{vmatrix} 3 & 1 & x \\ -1 & 3 & 4 \\ x & 1 & 0 \end{vmatrix} = -30$

iii. $\begin{vmatrix} x-2 & 1 \\ 5 & x+2 \end{vmatrix} = 0$

Q.6. If $A = \begin{vmatrix} 1 & -1 & 2 \\ 3 & 2 & 5 \\ -1 & 0 & 4 \end{vmatrix}$ and $B = \begin{vmatrix} 2 & 1 & -1 \\ 1 & 3 & 4 \\ -1 & 2 & 1 \end{vmatrix}$, then find $A - B$.

Q.7. Write the minor and cofactor of 3 and 4 in the matrix $\begin{bmatrix} 3 & 1 & -4 \\ 2 & 5 & 6 \\ 1 & 4 & 8 \end{bmatrix}$.

Q.8. Solve the following system of equations using Cramer's Rule:

i.
$$3x - 4y = -2$$
$$x + y = 6$$

ii.
$$x - y = 2$$
$$x + 4y = 5$$

iii. $2x + 2y + z = 1$
$$x - y + 6z = 21$$
$$3x + 2y - z = -4$$

iv. $x - 2y + z = -1$
$$3x + y - 2z = 4$$
$$y - z = 1$$
$$x + y + z = 0$$

v. $2x - y - 4z = 15$
$$x - 2y - z = 7$$

ANSWERS TO MULTIPLE-CHOICE QUESTIONS

Answer 1: (b)
Answer 2: (a)

Answer 3: (c)
Answer 4: (c)
Answer 5: (d)
Answer 6: (a)

ANSWERS TO EXERCISE 2

Answer 1:

   i. 1000
   ii. 0
   iii. −41
   iv. 9

Answer 5:

   i. x = 3, 4
   ii. x = −2, 3
   iii. x = ±3

Answer 6: $\begin{vmatrix} -1 & -2 & 3 \\ 2 & -1 & 1 \\ 0 & -2 & 3 \end{vmatrix}$

Answer 7: Minor of 3 = 16 and cofactor of 3 = 16

   Minor of 4 = 26 and cofactor of 4 = −26

Answer 8:

   i. $\left\{ \left(\dfrac{22}{7}, \dfrac{20}{7} \right) \right\}$

   ii. $\left\{ \left(\dfrac{13}{5}, \dfrac{3}{5} \right) \right\}$

   iii. $\{(1,-2,3)\}$

   iv. $\{(1,1,0)\}$

   v. $\{(3,-1,-2)\}$

3 More about Matrices

3.1 INTRODUCTION

In Chapter 1, we discussed various types of matrices; to know more about eigen values and eigen vectors, to solve linear equations using matrices, etc., we need to learn special types of matrices which include transpose, symmetric, skew symmetric, singular, non-singular, adjoint and inverse of matrices, which are discussed in this chapter in detail.

3.2 SPECIAL MATRICES

1. Transpose of Matrix

   Let A be a m × n matrix and A = [a_{ij}]. The matrix obtained by interchanging rows and columns of A is called the *transpose* of A. It is denoted by A^T or A'. A^T is of order n × m.

   For example, if $A = \begin{bmatrix} 1 & 2 & 3 & 4 \\ 5 & 6 & 7 & 8 \\ 9 & 0 & -1 & -2 \end{bmatrix}$, then $A^T = \begin{bmatrix} 1 & 5 & 9 \\ 2 & 6 & 0 \\ 3 & 7 & -1 \\ 4 & 8 & -2 \end{bmatrix}$.

2. Symmetric Matrix

   A square matrix A is called *symmetric* if $A = A^T$.

   For example, if $A = \begin{bmatrix} 1 & 2 & 3 \\ 2 & 6 & 7 \\ 3 & 7 & -1 \end{bmatrix}$, then $A^T = \begin{bmatrix} 1 & 2 & 3 \\ 2 & 6 & 7 \\ 3 & 7 & -1 \end{bmatrix} = A$.

   Hence, A is symmetric.

3. Skew Symmetric Matrix

   A square matrix A is called *skew symmetric* if $A = -A^{-T}$.

   For example, if $A = \begin{bmatrix} 0 & -5 & 2 \\ 5 & 0 & -4 \\ -2 & 4 & 0 \end{bmatrix}$, then

   $A^T = \begin{bmatrix} 0 & 5 & -2 \\ -5 & 0 & 4 \\ 2 & -4 & 0 \end{bmatrix} = (-1)A$.

   Hence, A is skew symmetric.

4. Singular and Non-Singular Matrices

   A square matrix A is called *singular* if $|A| = 0$ and is *non-singular* if $|A| \neq 0$.

For example, if $A = \begin{bmatrix} 1 & 2 \\ 3 & 6 \end{bmatrix}$, then $|A| = 6 - 6 = 0$. Hence, A is singular.

If $A = \begin{bmatrix} 1 & 2 \\ 3 & 8 \end{bmatrix}$, then $|A| = 8 - 6 = 2 \neq 0$. Hence, A is non-singular.

Illustration 3.1: If $A = \begin{bmatrix} x-2 & 1 \\ 5 & x+2 \end{bmatrix}$ is singular, then find x.

Solution: Here, as A is singular, $\begin{vmatrix} x-2 & 1 \\ 5 & x+2 \end{vmatrix} = 0,$

$$\Rightarrow (x-2)(x+2) - 5 = 0$$

$$\Rightarrow x^2 - 9 = 0$$

$$\Rightarrow x^2 = 9$$

$$\Rightarrow x = \pm 3$$

5. Adjoint of Matrix

Let $A = (a_{ij})$ be a square matrix of order n and (c_{ij}) be a matrix obtained by replacing each element a_{ij} by its corresponding cofactor c_{ij}, then $(c_{ij})^T$ is called the *adjoint of* A. It is denoted as adjA.

For example, if $A = \begin{bmatrix} 1 & 0 & -1 \\ 1 & 3 & 1 \\ 0 & 1 & 2 \end{bmatrix}$, then cofactors of A are:

Cofactor of $1 = A_{11} = 5$, $A_{12} = (-2)$, $A_{13} = 1$,
$A_{21} = (-1)$, $A_{22} = 2$, $A_{23} = (-1)$,
$A_{31} = 3$, $A_{32} = (-2)$, $A_{33} = 3$.

Hence, the matrix of cofactors is $C = \begin{bmatrix} 5 & -2 & 1 \\ -1 & 2 & -1 \\ 3 & -2 & 3 \end{bmatrix}.$

So $C^T = \begin{bmatrix} 5 & -1 & 3 \\ -2 & 2 & -2 \\ 1 & -1 & 3 \end{bmatrix}.$

Thus, adjA $= C^T = \begin{bmatrix} 5 & -1 & 3 \\ -2 & 2 & -2 \\ 1 & -1 & 3 \end{bmatrix}.$

Note: The adjoint of a 2 × 2 matrix $A = \begin{bmatrix} a & b \\ c & d \end{bmatrix}$ is defined as

$adjA = \begin{bmatrix} d & -b \\ -c & a \end{bmatrix}$.

6. Inverse of Matrix

If A is a non-singular square matrix, then $A^{-1} = \dfrac{adjA}{|A|}$ is the *inverse of*

the matrix.

For example, if $A = \begin{bmatrix} 3 & 4 \\ 1 & 2 \end{bmatrix}$, then $adjA = \begin{bmatrix} 2 & -4 \\ -1 & 3 \end{bmatrix}$ and $|A| = 2$, hence

$A^{-1} = \dfrac{1}{2} \begin{bmatrix} 2 & -4 \\ -1 & 3 \end{bmatrix}$.

Alternatively, for a non-singular matrix A of order n × n, if there exists another matrix B of order n × n such that the product is the identity matrix I of order n × n, i.e. AB = BA = I, then B is said to be the inverse of A and is written as $B = A^{-1}$.

Note: If A is singular then the solution does not exist, and if A is non-singular then the solution exists.

Illustration 3.2: If $A = \begin{bmatrix} 1 & -3 \\ -2 & 7 \end{bmatrix}$ and $B = \begin{bmatrix} 7 & 3 \\ 2 & 1 \end{bmatrix}$, then show that B is the inverse

of A.

Solution: $AB = \begin{bmatrix} 1 & -3 \\ -2 & 7 \end{bmatrix}\begin{bmatrix} 7 & 3 \\ 2 & 1 \end{bmatrix} = \begin{bmatrix} 1 & 0 \\ 0 & 1 \end{bmatrix}$

$BA = \begin{bmatrix} 7 & 3 \\ 2 & 1 \end{bmatrix}\begin{bmatrix} 1 & -3 \\ -2 & 7 \end{bmatrix} = \begin{bmatrix} 1 & 0 \\ 0 & 1 \end{bmatrix}$

Hence, AB = BA = I and $B = A^{-1} = \begin{bmatrix} 7 & 3 \\ 2 & 1 \end{bmatrix}$.

Illustration 3.3: If it exists, find the inverse of matrix $A = \begin{bmatrix} 0 & 1 & 2 \\ 1 & 2 & 3 \\ 3 & 1 & 1 \end{bmatrix}$.

Solution: Here, $A = \begin{bmatrix} 0 & 1 & 2 \\ 1 & 2 & 3 \\ 3 & 1 & 1 \end{bmatrix}$

$\therefore |A| = 0 + 9 + 2 - 12 - 1 - 0 = -2 \neq 0$

Hence, A is non-singular, so a solution exists.

The cofactors of $0 = A_{11} = (-1)$, $A_{12} = 8$, $A_{13} = (-5)$, $A_{21} = 1$, $A_{22} = (-6)$, $A_{23} = 3$, $A_{31} = (-1)$, $A_{32} = 2$, $A_{33} = (-1)$.

The matrix of cofactors is $C = \begin{bmatrix} -1 & 8 & -5 \\ 1 & -6 & 3 \\ -1 & 2 & -1 \end{bmatrix}$.

So $C^T = \begin{bmatrix} -1 & 1 & -1 \\ 8 & -6 & 2 \\ -5 & 3 & -1 \end{bmatrix}$.

Thus, $\text{adj} A = C^T = \begin{bmatrix} -1 & 1 & -1 \\ 8 & -6 & 2 \\ -5 & 3 & -1 \end{bmatrix}$.

$$\therefore A^{-1} = \frac{1}{-2} \begin{bmatrix} -1 & 1 & -1 \\ 8 & -6 & 2 \\ -5 & 3 & -1 \end{bmatrix}.$$

3.3 SOLUTION OF LINEAR EQUATIONS BY MATRICES

Let us consider the linear system:

$$a_{11}x_1 + a_{12}x_2 + \dots + a_{1n}x_n = b_1$$

$$a_{21}x_1 + a_{22}x_2 + \dots + a_{2n}x_n = b_2$$

$$\vdots \qquad \vdots \qquad \qquad \vdots$$

$$a_{n1}x_1 + a_{n2}x_2 + \dots + a_{nn}x_n = b_n$$

(3.1)

In matrix form, it can be written as:

$$\begin{bmatrix} a_{11} & a_{12} & \dots & a_{1n} \\ a_{21} & a_{22} & \dots & a_{2n} \\ \vdots & \vdots & \vdots & \vdots \\ a_{n1} & a_{n2} & \dots & a_{nn} \end{bmatrix} \begin{bmatrix} x_1 \\ x_2 \\ \vdots \\ x_n \end{bmatrix} = \begin{bmatrix} b_1 \\ b_2 \\ \vdots \\ b_n \end{bmatrix}$$

Let $A = \begin{bmatrix} a_{11} & a_{12} & \dots & a_{1n} \\ a_{21} & a_{22} & \dots & a_{2n} \\ \vdots & \vdots & \vdots & \vdots \\ a_{n1} & a_{n2} & \dots & a_{nn} \end{bmatrix}$, $X = \begin{bmatrix} x_1 \\ x_2 \\ \vdots \\ x_n \end{bmatrix}$, $B = \begin{bmatrix} b_1 \\ b_2 \\ \vdots \\ b_n \end{bmatrix}$.

Then the above equation can be written as $AX = B$.

If $B = 0$ in equation (3.1), then it is called a system of homogenous linear equations, and if $B \neq 0$, it is called a system of non-homogenous linear equations.

Now, if $B \neq 0$ and A is non-singular, then A^{-1} exists.

Multiplying both sides of the equation $AX = B$ by A^{-1}, we get

$$\therefore A^{-1}(AX) = A^{-1}B$$

$\therefore X = A^{-1}B$; where $A^{-1}B$ is an $n \times 1$ column matrix. As X and $A^{-1}B$ are equal, each element in X is equal to that element in $A^{-1}B$. Elements of X constitute the solution of linear equations. If A is singular, then the inverse does not exist and the system has no solution.

Illustration 3.4: Find the solution set of the following using matrices:

$$4x + 8y + z = -6$$

$$2x - 3y + 2z = 0$$

$$x + 7y - 3z = -8$$

Solution: Here, $A = \begin{bmatrix} 4 & 8 & 1 \\ 2 & -3 & 2 \\ 1 & 7 & -3 \end{bmatrix}$

$$\therefore |A| = 61 \neq 0$$

So A^{-1} exists.

And we know that $A^{-1} = \dfrac{1}{|A|} \text{adj} A$

where $\text{adj} A = \begin{bmatrix} -5 & 31 & 19 \\ 8 & -13 & -6 \\ 17 & -20 & -28 \end{bmatrix}$.

$$\therefore A^{-1} = \frac{1}{61}\begin{bmatrix} -5 & 31 & 19 \\ 8 & -13 & -6 \\ 17 & -20 & -28 \end{bmatrix}$$

As $X = A^{-1}B$,

$$\therefore \begin{bmatrix} x \\ y \\ z \end{bmatrix} = \frac{1}{61}\begin{bmatrix} -5 & 31 & 19 \\ 8 & -13 & -6 \\ 17 & -20 & -28 \end{bmatrix}\begin{bmatrix} -6 \\ 0 \\ -8 \end{bmatrix}$$

$$\therefore \begin{bmatrix} x \\ y \\ z \end{bmatrix} = \frac{1}{61}\begin{bmatrix} -122 \\ 0 \\ 122 \end{bmatrix} = \begin{bmatrix} -2 \\ 0 \\ 2 \end{bmatrix}$$

Hence, the solution set is $\{(x, y, z)\} = \{(-2, 0, 2)\}$.

Illustration 3.4: Find the solution set of the following using matrices:

$$x + y - 2z = 3$$

$$3x - y + z = 5$$

$$3x + 3y - 6z = 9$$

Solution: Here, $A = \begin{bmatrix} 1 & 1 & -2 \\ 3 & -1 & 1 \\ 3 & 3 & -6 \end{bmatrix}$

$$\therefore |A| = 0$$

Hence, the solution of the given linear equations does not exist.

3.4 EIGEN VALUES AND EIGEN VECTORS

Let $A = [a_{ij}]$ be of order n × n and λ be a scalar. If for a non-zero column matrix (n × 1) X, $AX = \lambda X$, then X is called the *eigen vector* of A corresponding to λ and λ is called the *eigen value* of A.

Note: If the eigen value of the matrix exists, then it is unique.

Proof: If possible, let us suppose that λ_1 and λ_2 are the two eigen values corresponding to the eigen vector X. Thus, we have $AX = \lambda_1 X$ and $AX = \lambda_2 X$ which implies that $\lambda_1 X = \lambda_2 X$. But as X is a non-zero matrix, hence $\lambda_1 = \lambda_2$. Let $X = \begin{bmatrix} x_1 \\ x_2 \\ \vdots \\ x_n \end{bmatrix}$.

The equation $AX = \lambda X$ can also be expressed as

$$\begin{bmatrix} a_{11} - \lambda & a_{12} & \cdots & a_{1n} \\ a_{21} & a_{22} - \lambda & \cdots & a_{2n} \\ \vdots & \vdots & \cdots & \vdots \\ a_{n1} & a_{n2} & \cdots & a_{nn} - \lambda \end{bmatrix} \begin{bmatrix} x_1 \\ x_2 \\ \vdots \\ x_n \end{bmatrix} = \begin{bmatrix} 0 \\ 0 \\ \vdots \\ 0 \end{bmatrix}$$

i.e. $\left[A - \lambda I_n \right] X = 0$

$$\therefore \left(a_{11} - \lambda \right) x_1 + a_{12} x_2 + \cdots + a_{1n} x_n = 0,$$

$$a_{21} x_1 + \left(a_{22} - \lambda \right) x_2 + \cdots + a_{2n} x_n = 0,$$

$$\cdots \quad \cdots \quad \cdots \quad \cdots \quad \cdots \quad \cdots \quad \cdots \quad \cdots \quad \cdots$$

$$\cdots \quad \cdots \quad \cdots \quad \cdots \quad \cdots \quad \cdots \quad \cdots \quad \cdots \quad \cdots$$

$$a_{n1} x_1 + a_{2n} x_n + \cdots + \left(a_{nn} - \lambda \right) x_n = 0$$

The necessary and sufficient condition for the existence of a non-zero solution of the above equation is that the matrix $A - \lambda I_n$ is singular. Hence,

$$\left| A - \lambda I_n \right| \tag{3.2}$$

where equation (3.2) is called the *characteristic equation* of the matrix A, and this equation is of degree n in λ; hence its n roots give us the eigen values of A. Substituting these values in the equation $\left[A - \lambda I_n \right] X = 0$ yields the corresponding eigen vectors of A.

Illustration 3.5: Find the eigen values and eigen vectors of the matrix $A = \begin{bmatrix} 1 & 4 \\ 3 & 2 \end{bmatrix}$.

Solution: Here, $A = \begin{bmatrix} 1 & 4 \\ 3 & 2 \end{bmatrix}$

So the characteristic equation of A is $\begin{vmatrix} 1-\lambda & 4 \\ 3 & 2-\lambda \end{vmatrix} = 0.$

$$\therefore \lambda^2 - 3\lambda - 10 = 0$$

$$\therefore (\lambda - 5)(\lambda + 2) = 0$$

Hence, $\lambda = 5$ and $\lambda = -2$ are the eigen values.

For $\lambda = 5$, the equation $\left[A - \lambda I_2 \right] X = 0$ becomes $\begin{bmatrix} -4 & 4 \\ 3 & -3 \end{bmatrix} \begin{bmatrix} x_1 \\ x_2 \end{bmatrix} = \begin{bmatrix} 0 \\ 0 \end{bmatrix}$

$$\Rightarrow -4x_1 + 4x_2 = 0$$

$$3x_1 - 3x_2 = 0$$

$$\Rightarrow x_1 = x_2$$

Hence, the eigen vector corresponding to $\lambda = 5$ is $a \begin{bmatrix} 1 \\ 1 \end{bmatrix}$, where a is any real number.

For $\lambda = -2$, the equation $\left[A - \lambda I_2 \right] X = 0$ becomes $\begin{bmatrix} 3 & 4 \\ 3 & 4 \end{bmatrix} \begin{bmatrix} x_1 \\ x_2 \end{bmatrix} = \begin{bmatrix} 0 \\ 0 \end{bmatrix}$

$$\Rightarrow 3x_1 + 4x_2 = 0$$

$$3x_1 + 4x_2 = 0$$

$$\Rightarrow x_1 = \frac{-4}{3} x_2$$

Hence, the eigen vector corresponding to $\lambda = -2$ is $a \begin{bmatrix} -4 \\ 3 \\ 1 \end{bmatrix}$, where a is any real

number.

Remark:

i. The eigen value of the identity matrix is 1.
ii. The eigen values of a diagonal matrix are its diagonal elements.
iii. The eigen values of a matrix are the same as the eigen values of its transpose.

Illustration 3.6: Find eigen values and eigen vectors of the matrix $A = \begin{bmatrix} 2 & 1 & 0 \\ 0 & 1 & -1 \\ 0 & 2 & 4 \end{bmatrix}$

Solution: Here $A = \begin{bmatrix} 2 & 1 & 0 \\ 0 & 1 & -1 \\ 0 & 2 & 4 \end{bmatrix}$

So the characteristic equation of A is $\begin{vmatrix} 2-\lambda & 1 & 0 \\ 0 & 1-\lambda & -1 \\ 0 & 2 & 4-\lambda \end{vmatrix} = 0$

$$\therefore (\lambda - 2)^2 (\lambda - 3) = 0$$

Hence, $\lambda = 2$ and $\lambda = 3$ are the eigen values of A.

It is not necessary that all the roots of the characteristic equation be distinct. Thus, here $\lambda = 2$ is a repeated root.

For $\lambda = 2$, the equation $\left[A - \lambda I_3 \right] X = 0$ becomes $\begin{bmatrix} 0 & 1 & 0 \\ 0 & -1 & -1 \\ 0 & 2 & 2 \end{bmatrix} \begin{bmatrix} x_1 \\ x_2 \\ x_3 \end{bmatrix} = \begin{bmatrix} 0 \\ 0 \\ 0 \end{bmatrix}$

$$\Rightarrow x_2 = 0$$

$$-x_2 - x_3 = 0$$

$$2x_2 + 2x_3 = 0$$

Thus, for $\lambda = 2$, we have $x_2 = x_3 = 0$ and x_1 arbitrary by solving the above system of equations.

Hence, the eigen vector corresponding to $\lambda = 2$ is $a \begin{bmatrix} 1 \\ 0 \\ 0 \end{bmatrix}$, where a is any real

number.

For $\lambda = 3$, the equation $\left[A - \lambda I_3\right]X = 0$ becomes $\begin{bmatrix} -1 & 1 & 0 \\ 0 & -2 & -1 \\ 0 & 2 & 1 \end{bmatrix}\begin{bmatrix} x_1 \\ x_2 \\ x_3 \end{bmatrix} = \begin{bmatrix} 0 \\ 0 \\ 0 \end{bmatrix}$

$$\Rightarrow -x_1 + x_2 = 0$$

$$-2x_2 - x_3 = 0$$

$$2x_2 + x_3 = 0$$

Thus, for $\lambda = 3$, we have $x_1 = x_2$ and $x_3 = -2x_2$.

Hence, the eigen vector corresponding to $\lambda = 3$ is $a\begin{bmatrix} 1 \\ 1 \\ -2 \end{bmatrix}$, where a is any real

number.

Illustration 3.7: Find eigen values and eigen vectors of the matrix
$A = \begin{bmatrix} 0 & -2 & -2 \\ -2 & -3 & -2 \\ 3 & 6 & 5 \end{bmatrix}$

Solution: Here, $A = \begin{bmatrix} 0 & -2 & -2 \\ -2 & -3 & -2 \\ 3 & 6 & 5 \end{bmatrix}$

So the characteristic equation of A is $\begin{vmatrix} 0-\lambda & -2 & -2 \\ -2 & -3-\lambda & -2 \\ 3 & 6 & 5-\lambda \end{vmatrix} = 0$

$$\therefore \lambda^3 - 2\lambda^2 - \lambda + 2 = 0$$

As $\lambda = 1, \lambda = -1, \lambda = 2$ are the roots of the above equation, $\lambda = 1, \lambda = -1$ and $\lambda = 2$ are the eigen values of A.

For $\lambda = 1$, the equation $\left[A - \lambda I_3\right]X = 0$ becomes $\begin{bmatrix} -1 & -2 & -2 \\ -2 & -4 & -2 \\ 3 & 6 & 4 \end{bmatrix}\begin{bmatrix} x_1 \\ x_2 \\ x_3 \end{bmatrix} = \begin{bmatrix} 0 \\ 0 \\ 0 \end{bmatrix}$

$$\Rightarrow x_1 + 2x_2 + 2x_3 = 0$$

$$x_1 + 2x_2 + x_3 = 0$$

$$3x_1 + 6x_2 + 4x_3 = 0$$

Thus, for $\lambda = 1$, we have $x_1 = -2x_2$ and $x_3 = 0$ on solving the above system of equations.

Hence, the eigen vector corresponding to $\lambda = 1$ is $a \begin{bmatrix} -2 \\ 1 \\ 0 \end{bmatrix}$, where a is any real number.

For $\lambda = -1$, the equation $\left[A - \lambda I_3\right]X = 0$ becomes $\begin{bmatrix} 1 & -2 & -2 \\ -2 & -2 & -2 \\ 3 & 6 & 6 \end{bmatrix}\begin{bmatrix} x_1 \\ x_2 \\ x_3 \end{bmatrix} = \begin{bmatrix} 0 \\ 0 \\ 0 \end{bmatrix}$

$$\Rightarrow x_1 - 2x_2 - 2x_3 = 0$$

$$-2x_1 - 2x_2 - 2x_3 = 0$$

$$3x_1 + 6x_2 + 6x_3 = 0$$

Thus, for $\lambda = 1$, we have $x_1 = 0$ and $x_2 = -x_3$.

Hence, the eigen vector corresponding to $\lambda = 1$ is $a \begin{bmatrix} 0 \\ 1 \\ -1 \end{bmatrix}$, where a is any real number.

For $\lambda = 2$, the equation $\left[A - \lambda I_3\right]X = 0$ becomes $\begin{bmatrix} -2 & -2 & -2 \\ -2 & -5 & -2 \\ 3 & 6 & 3 \end{bmatrix}\begin{bmatrix} x_1 \\ x_2 \\ x_3 \end{bmatrix} = \begin{bmatrix} 0 \\ 0 \\ 0 \end{bmatrix}$

$$\Rightarrow -2x_1 - 2x_2 - 2x_3 = 0$$

$$-2x_1 - 5x_2 - 2x_3 = 0$$

$$3x_1 + 6x_2 + 3x_3 = 0$$

Thus, for $\lambda = 2$, we have $x_2 = 0$ and $x_1 = -x_3$.

Hence, the eigen vector corresponding to $\lambda = 2$ is $a \begin{bmatrix} 1 \\ 0 \\ -1 \end{bmatrix}$, where a is any real number.

3.5   PROPERTIES OF EIGEN VALUES AND EIGEN VECTORS

   i. An $n \times n$ matrix has at least one eigen value and at most n numerically different eigen values.
   ii. If A is triangular (either upper triangular or lower triangular) matrix, then the diagonal elements of A are the eigen values of A, where A is an $n \times n$ invertible matrix.

iii. If λ is an eigen value of x with eigen vector x, then $\dfrac{1}{\lambda}$ is the eigen value of A^{-1} with eigen vector x, where A is an n × n invertible matrix.

iv. If λ is an eigen value of A, then λ is an eigen value of A^T, where A is an n × n invertible matrix.

v. The sum of eigen values of A is equal to tr(A), where tr(A) denotes the trace of A and A is an n × n invertible matrix.

vi. The product of the eigen values of A is equal to the determinant of A, where A is an n × n invertible matrix.

vii. The eigen values of a square matrix A are the roots of the characteristic equation of A.

viii. For a real square matrix A, if $\alpha + i\beta$ is an eigen value, then its conjugate $\alpha - i\beta$ is also an eigen value. When A is complex, this property does not hold.

Illustration 3.8: If λ is an eigen value of an invertible matrix $A = \left[a_{ij} \right]_n$, then show that

i. $\dfrac{1}{\lambda}$ is the eigen value of A^{-1}.

ii. $\dfrac{|A|}{\lambda}$ is the eigen value of adjA.

Solution:

i. λ is an eigen value of A. Therefore, there exists a non-zero column matrix X such that

$$AX = \lambda X$$

$$\therefore X = A^{-1}\lambda X = \lambda A^{-1}X$$

$$\therefore \frac{1}{\lambda}X = A^{-1}X$$

which shows that $\dfrac{1}{\lambda}$ is the eigen value of A^{-1}.

ii. Similarly, here also
$AX = \lambda X$ (column matrix exists)

$$\left(adjA \right)\left(AX \right) = \left(adjA \right)\lambda X$$

$$\Rightarrow \left(adjA.A \right)X = \lambda \left(adjA \right)X$$

$$\Rightarrow |A| X = \lambda (\text{adj} A) X$$

$$\Rightarrow (\text{adj} A) X = \frac{|A|}{\lambda} X$$

Hence, $\dfrac{|A|}{\lambda}$ is the eigen value of adjA.

3.6   CAYLEY–HAMILTON THEOREM

Theorem: Every square matrix $A = \left[a_{ij} \right]_n$ satisfies its own characteristic equation

$$\text{i.e. If } |A - \lambda I_n| = (-1)^n \lambda^n + k_1 \lambda^{n-1} + k_2 \lambda^{n-2} + \cdots + k_n = 0 \qquad (3.3)$$

is the characteristic equation of A, then

$$(-1)^n A^n + k_1 A^{n-1} + k_2 A^{n-2} + \cdots + k_n I = 0 \qquad (3.4)$$

Proof: Let $B = \text{adj}(A - \lambda I_n)$

Each element of the matrix $A - \lambda I_n$ is, at the most, degree 1 in λ. Therefore the elements of B are polynomials in λ of, at the most, degree $n - 1$. Thus, we can express matrix B in the form

$$B = B_1 \lambda^{n-1} + B_2 \lambda^{n-2} + \cdots + B_{n-2} \lambda + B_{n-1} \qquad (3.5)$$

Where $B_1, B_2, \cdots, B_{n-1}$ are $n \times n$ matrices.

We know that for any square matrix C, $C^{-1} = \dfrac{1}{|C|} \text{adj} C$.

Hence, $C(\text{adj} C) = |C| I_n$.

Thus, $\left[A - \lambda I_n \right] B = |A - \lambda I_n| \cdot I_n$.

From equations (3.3) and (3.5) we have,

$\left[A - \lambda I_n \right]$

$$\left[B_1 \lambda^{n-1} + B_2 \lambda^{n-2} + \cdots + B_{n-2} \lambda + B_{n-1} \right] = \left[(-1)^n \lambda^n + k_1 \lambda^{n-1} + k_2 \lambda^{n-2} + \cdots + k_n \right] I_n$$

Comparing the powers of λ from both sides of the above equation, we get,

$$-B_1 = (-1)^n I_n$$

$$AB_1 - B_2 = k_1 I_n$$

$$AB_2 - B_3 = k_2 I_n$$

$$\cdots \quad\quad \cdots \quad\quad \cdots$$

$$\cdots \quad\quad \cdots \quad\quad \cdots$$

$$AB_{n-2} - B_{n-1} = k_{n-1} I_n$$

$$AB_{n-1} = k_n I_n$$

Multiplying the above equations by $A^n, A^{n-1}, A^{n-2}, \cdots, A, I_n$, respectively, and adding, we get,

$$(-1)^n A^n + k_1 A^{n-1} + k_2 A^{n-2} + \cdots + k_{n-1} A + k_n I_n = 0 \qquad (3.6)$$

which proves Cayley–Hamilton theorem.

Note: Multiplying equation (3.6) by A^{-1}, we get,

$$A^{-1} = \frac{-1}{k_n}\left[(-1)^n A^{n-1} + k_1 A^{n-2} + \cdots + k_{n-1} I_n\right]$$

Hence, we can find the inverse of an invertible matrix using Cayley–Hamilton theorem.

Illustration 3.9: Find the characteristic equation of the matrix $A = \begin{bmatrix} 1 & 1 & 3 \\ 1 & 3 & -3 \\ -2 & -4 & -4 \end{bmatrix}$

and verify Cayley–Hamilton theorem. Also, find A^{-1}.

Solution: Here, $A = \begin{bmatrix} 1 & 1 & 3 \\ 1 & 3 & -3 \\ -2 & -4 & -4 \end{bmatrix}$

So the characteristic equation of the matrix A is $\begin{vmatrix} 1-\lambda & 1 & 3 \\ 1 & 3-\lambda & -3 \\ -2 & -4 & -4-\lambda \end{vmatrix} = 0$

On expansion we get, $(1-\lambda)\left[\lambda^2 + \lambda - 24\right] - \left[-10 - \lambda\right] + 3\left[2 - 2\lambda\right] = 0$.

On simplification we get, $-\lambda^3 + 20\lambda - 8 = 0$.

Hence, $\lambda^3 - 20\lambda + 8 = 0$ is the required characteristic equation.

Now, $A^2 = A \cdot A$

$$= \begin{bmatrix} 1 & 1 & 3 \\ 1 & 3 & -3 \\ -2 & -4 & -4 \end{bmatrix} \begin{bmatrix} 1 & 1 & 3 \\ 1 & 3 & -3 \\ -2 & -4 & -4 \end{bmatrix}$$

$$= \begin{bmatrix} -4 & -8 & -12 \\ 10 & 22 & 6 \\ 2 & 2 & 22 \end{bmatrix}$$

$$A^3 = A^2 \cdot A$$

$$= \begin{bmatrix} -4 & -8 & -12 \\ 10 & 22 & 6 \\ 2 & 2 & 22 \end{bmatrix} \begin{bmatrix} 1 & 1 & 3 \\ 1 & 3 & -3 \\ -2 & -4 & -4 \end{bmatrix}$$

$$= \begin{bmatrix} 12 & 20 & 60 \\ 20 & 52 & -60 \\ -40 & -80 & -88 \end{bmatrix}$$

$$-20A = \begin{bmatrix} -20 & -20 & -60 \\ -20 & -60 & 60 \\ 40 & 80 & 80 \end{bmatrix}$$

$$8I_3 = \begin{bmatrix} 8 & 0 & 0 \\ 0 & 8 & 0 \\ 0 & 0 & 8 \end{bmatrix}$$

Obviously, it can be seen that $A^3 - 20A + 8I_3 = 0$.
So Cayley–Hamilton theorem is verified.
Multiplying by A^{-1} we get, $A^2 - 20I + 8A^{-1} = 0$

$$\Rightarrow 8A^{-1} = 20I - A^2$$

$$\Rightarrow A^{-1} = \frac{1}{8}\left[20I - A^2\right]$$

$$\Rightarrow A^{-1} = \frac{1}{8}\left[20I - A^2\right]$$

$$\Rightarrow A^{-1} = \frac{1}{8}\begin{bmatrix} 24 & 8 & 12 \\ -10 & -2 & -6 \\ -2 & -2 & -2 \end{bmatrix}$$

Illustration 3.10: For the matrix $A = \begin{bmatrix} 2 & -1 & 1 \\ -1 & 2 & -1 \\ 1 & -1 & 2 \end{bmatrix}$, verify Cayley–Hamilton theorem. Also, find A^{-1}.

Solution: Here, $A = \begin{bmatrix} 2 & -1 & 1 \\ -1 & 2 & -1 \\ 1 & -1 & 2 \end{bmatrix}$.

So the characteristic equation of the matrix A is $\begin{vmatrix} 2-\lambda & -1 & 1 \\ -1 & 2-\lambda & -1 \\ 1 & -1 & 2-\lambda \end{vmatrix} = 0$.

On expansion we get, $(2-\lambda)\left[\lambda^2 - 4\lambda + 3\right] + \left[-1+\lambda\right] + 1\left[-1+\lambda\right] = 0$.

On simplification we get, $-\lambda^3 + 6\lambda^2 - 9\lambda + 4 = 0$.

Hence, $\lambda^3 - 6\lambda^2 + 9\lambda - 4 = 0$ is the required characteristic equation.

Now, for the verification of Cayley–Hamilton theorem, we need to show that $A^3 - 6A^2 + 9A - 4I_3 = 0$.

Now, $A^2 = \begin{bmatrix} 2 & -1 & 1 \\ -1 & 2 & -1 \\ 1 & -1 & 2 \end{bmatrix}\begin{bmatrix} 2 & -1 & 1 \\ -1 & 2 & -1 \\ 1 & -1 & 2 \end{bmatrix}$

$$= \begin{bmatrix} 6 & -5 & 5 \\ -5 & 6 & -5 \\ 5 & -5 & 6 \end{bmatrix}$$

And, $A^3 = \begin{bmatrix} 6 & -5 & 5 \\ -5 & 6 & -5 \\ 5 & -5 & 6 \end{bmatrix}\begin{bmatrix} 2 & -1 & 1 \\ -1 & 2 & -1 \\ 1 & -1 & 2 \end{bmatrix}$

$$= \begin{bmatrix} 22 & -21 & 21 \\ -21 & 22 & -21 \\ 21 & -21 & 22 \end{bmatrix}$$

So $A^3 - 6A^2 + 9A - 4I_3$

$$= \begin{bmatrix} 22 & -21 & 21 \\ -21 & 22 & -21 \\ 21 & -21 & 22 \end{bmatrix} - 6\begin{bmatrix} 6 & -5 & 5 \\ -5 & 6 & -5 \\ 5 & -5 & 6 \end{bmatrix} + 9\begin{bmatrix} 2 & -1 & 1 \\ -1 & 2 & -1 \\ 1 & -1 & 2 \end{bmatrix} - 4\begin{bmatrix} 1 & 0 & 0 \\ 0 & 1 & 0 \\ 0 & 0 & 1 \end{bmatrix}$$

$$= \begin{bmatrix} 0 & 0 & 0 \\ 0 & 0 & 0 \\ 0 & 0 & 0 \end{bmatrix}$$

$$= 0$$

Hence, $A^3 - 6A^2 + 9A - 4I_3 = 0$

$$\Rightarrow A^2 - 6A + 9I - 4A^{-1} = 0$$

$$\Rightarrow 4A^{-1} = A^2 - 6A + 9I$$

$$\Rightarrow A^{-1} = \frac{1}{4}\left[A^2 - 6A + 9I\right]$$

$$\Rightarrow A^{-1} = \frac{1}{4}\left[\begin{bmatrix} 6 & -5 & 5 \\ -5 & 6 & -5 \\ 5 & -5 & 6 \end{bmatrix} - 6\begin{bmatrix} 2 & -1 & 1 \\ -1 & 2 & -1 \\ 1 & -1 & 2 \end{bmatrix} + 9\begin{bmatrix} 1 & 0 & 0 \\ 0 & 1 & 0 \\ 0 & 0 & 1 \end{bmatrix}\right]$$

$$\Rightarrow A^{-1} = \frac{1}{4}\begin{bmatrix} 3 & 1 & -1 \\ 1 & 3 & 1 \\ -1 & 1 & 3 \end{bmatrix}$$

MULTIPLE-CHOICE QUESTIONS

1. If $A = \left[a_{ij}\right]$ is of order m × n, then A^T is of order _____.

    (a)   m × n
    (b)   m × m
    (c)   n × n
    (d)   n × m

2. The value of x is _____ for the matrix $\begin{bmatrix} 2 & 3 \\ 6 & x \end{bmatrix}$ to be singular.

    (a)   6
    (b)   3
    (c)   8
    (d)   9

3. If A and B are symmetric then AB = _____.
    (a)   BA
    (b) $A^T B^T$

(c) $B^T A^T$

(d) Both (a) and (c)

4. The adjoint of a matrix $\begin{bmatrix} -2 & 3 \\ 6 & -1 \end{bmatrix}$ is ____.

(a) $\begin{bmatrix} -1 & 3 \\ 6 & -2 \end{bmatrix}$

(b) $\begin{bmatrix} -1 & -3 \\ -6 & -2 \end{bmatrix}$

(c) $\begin{bmatrix} 2 & 3 \\ 6 & 1 \end{bmatrix}$

(d) $\begin{bmatrix} 2 & -3 \\ -6 & 1 \end{bmatrix}$

5. The eigen value of the identity matrix is ____.
   (a) 1
   (b) 0
   (c) Both (a) and (b)
   (d) None of these

6. If for a given system of linear equations, the inverse does not exist, then the system has ____ solution.
   (a) Unique
   (b) No
   (c) More than one
   (d) None of these

EXERCISE 3

Q.1. Check which of the following matrices are singular or non-singular:

i. $\begin{bmatrix} 1 & 2 & 1 \\ 3 & 1 & -2 \\ 0 & 1 & -1 \end{bmatrix}$

ii. $\begin{bmatrix} 1 & 1 & -2 \\ 3 & -1 & 1 \\ 3 & 3 & -6 \end{bmatrix}$

iii. $\begin{bmatrix} 2 & 3 & -1 \\ 1 & 1 & 0 \\ 2 & -3 & 5 \end{bmatrix}$

Q.2. Check which of the following matrices are symmetric or skew symmetric:

i. $\begin{bmatrix} 0 & 3 & -5 \\ -3 & 0 & 6 \\ 5 & -6 & 0 \end{bmatrix}$

ii. $\begin{bmatrix} 2 & 6 & 7 \\ 6 & -2 & 3 \\ 7 & 3 & 0 \end{bmatrix}$

iii. $\begin{bmatrix} 1 & 2 & 3 \\ 2 & 4 & -3 \\ 3 & -3 & 6 \end{bmatrix}$

Q.3. Find k such that the following matrices are singular:

i. $\begin{bmatrix} k & 6 \\ 4 & 3 \end{bmatrix}$

ii. $\begin{bmatrix} 2 & 3 \\ 4 & k \end{bmatrix}$

iii. $\begin{bmatrix} 1 & 2 & -1 \\ -3 & 4 & k \\ -4 & 2 & 6 \end{bmatrix}$

iv. $\begin{bmatrix} 4 & k & 3 \\ 7 & 3 & 6 \\ 2 & 3 & 1 \end{bmatrix}$

Q.4. If exists, find the inverse of the following matrices:

i. $\begin{bmatrix} 1 & 3 \\ 2 & -1 \end{bmatrix}$

ii. $\begin{bmatrix} 2 & 1 \\ 6 & 3 \end{bmatrix}$

iii. $\begin{bmatrix} 5 & 3 \\ 1 & 1 \end{bmatrix}$

iv. $\begin{bmatrix} 0 & -2 & -3 \\ 1 & 3 & 3 \\ -1 & -2 & -2 \end{bmatrix}$

v. $\begin{bmatrix} 1 & 2 & -1 \\ -3 & 4 & 5 \\ -4 & 2 & 6 \end{bmatrix}$

Q.5. Using matrices, find the solution set of the following system:

i.
$$2x - 3y = -1$$
$$x + 4y = 5$$
$$x + y = 2$$

ii. $2x - z = 1$
$$2y - 3z = -1$$
$$x + y - 2z = 3$$

iii. $3x - y + z = 0$
$$3x + 3y - 6z = 8$$

Q.6. Find the eigen values and their corresponding eigen vectors of the following matrices:

i.
$$\begin{bmatrix} -3 & -6 & 7 \\ 1 & 1 & -1 \\ -4 & -6 & 8 \end{bmatrix}$$

ii.
$$\begin{bmatrix} 8 & -6 & 2 \\ -6 & 7 & -4 \\ 2 & -4 & 3 \end{bmatrix}$$

iii.
$$\begin{bmatrix} 5 & 3 & 6 \\ 0 & 1 & 0 \\ 0 & 4 & -2 \end{bmatrix}$$

Q.7. Show that the eigen values of the matrix $\begin{bmatrix} 1 & 1 & 1 \\ 1 & 1 & 1 \\ 1 & 1 & 1 \end{bmatrix}$ are 0 and 3 and find their eigen vectors.

Q.8. Prove that the eigen values of matrix $\begin{bmatrix} 9 & -7 & 7 \\ 3 & -1 & 3 \\ -5 & 5 & -3 \end{bmatrix}$ are 1 and 2. Also, prove that the eigen vector for the eigen values is $\begin{bmatrix} 7 \\ 3 \\ -5 \end{bmatrix}$ and the eigen vectors for the eigen value 2 are $a\begin{bmatrix} 0 \\ 1 \\ 1 \end{bmatrix}, a\begin{bmatrix} 1 \\ 1 \\ 0 \end{bmatrix}, a\begin{bmatrix} -1 \\ 0 \\ 1 \end{bmatrix}, a \neq 0.$

Q.9. Show that the eigen values of the matrix $\begin{bmatrix} 1 & 6 & 1 \\ 1 & 2 & 0 \\ 0 & 0 & 3 \end{bmatrix}$ are −1, 3 and 4.

Also show that the eigen vector corresponding to the eigen value 3 is

$a\begin{bmatrix} 1 \\ 1 \\ -4 \end{bmatrix}, a \neq 0.$

Q.10. Find the inverse of the following matrices using Cayley–Hamilton theorem:

i. $\begin{bmatrix} 2 & 3 \\ 3 & 5 \end{bmatrix}$

ii. $\begin{bmatrix} 7 & -1 & 3 \\ 6 & 1 & 4 \\ 2 & 4 & 8 \end{bmatrix}$

ANSWERS TO MULTIPLE-CHOICE QUESTIONS

Answer 1: (d)
Answer 2: (d)
Answer 3: (d)
Answer 4: (b)
Answer 5: (a)
Answer 6: (b)

ANSWERS TO EXERCISE 3

Answer 1:

  i. Non-singular
  ii. Singular
  iii. Singular

Answer 2:

  i. Skew symmetric
  ii. Symmetric
  iii. Symmetric

Answer 3:

  i. 8
  ii. 6

iii. 5
iv. 3

Answer 4:

i. $\begin{bmatrix} \dfrac{1}{7} & \dfrac{3}{7} \\ \dfrac{2}{7} & \dfrac{1}{-7} \end{bmatrix}$

ii. A^{-1} does not exist

iii. $\dfrac{1}{2}\begin{bmatrix} 1 & -3 \\ -1 & 5 \end{bmatrix}$

iv. $\begin{bmatrix} 0 & 2 & 3 \\ -1 & -3 & -3 \\ 1 & 2 & 2 \end{bmatrix}$

v. A^{-1} does not exist

Answer 5:

i. $\{(1,1)\}$

ii. $\{(1,1,1)\}$

iii. No solution

Answer 6:

i. $1,1,4;\ a\begin{bmatrix} 2 \\ 1 \\ 2 \end{bmatrix}, a \neq 0;\ a\begin{bmatrix} 1 \\ 0 \\ 1 \end{bmatrix}, a \neq 0$

ii. $0,3,15;\ a\begin{bmatrix} 1 \\ 2 \\ 2 \end{bmatrix}, a \neq 0;\ a\begin{bmatrix} 2 \\ 1 \\ -2 \end{bmatrix}, a \neq 0;\ a\begin{bmatrix} 2 \\ -2 \\ 1 \end{bmatrix}, a \neq 0$

iii. $1,-2,5;\ a\begin{bmatrix} \dfrac{-33}{4} \\ 3 \\ 4 \end{bmatrix}, a \neq 0;\ a\begin{bmatrix} 6 \\ 0 \\ -7 \end{bmatrix}, a \neq 0;\ a\begin{bmatrix} 1 \\ 0 \\ 0 \end{bmatrix}, a \neq 0$

Answer 7: The eigen vector corresponding to eigen value 3 is $a\begin{bmatrix}1\\1\\1\end{bmatrix}, a \neq 0$ and the

eigen vectors corresponding to eigen value 0 are $a\begin{bmatrix}0\\1\\-1\end{bmatrix}, a\begin{bmatrix}1\\0\\-1\end{bmatrix}, a\begin{bmatrix}1\\-1\\0\end{bmatrix}, a \neq 0$

Answer 10:

i. $A^{-1} = \begin{bmatrix} 5 & -3 \\ -3 & 2 \end{bmatrix}$

ii. $A^{-1} = \begin{bmatrix} \dfrac{-4}{25} & \dfrac{2}{5} & \dfrac{-7}{50} \\ \dfrac{-4}{5} & 1 & \dfrac{-1}{50} \\ \dfrac{11}{25} & \dfrac{-3}{5} & \dfrac{13}{50} \end{bmatrix}$

4 Application of Matrices and Determinants

4.1 INTRODUCTION TO THE APPLICATION OF MATRIX IN REAL LIFE

Matrix or matrices are most commonly used in mathematics, but other than that, where else can matrices be used? Even one will wonder from where the word matrix came from.

'Matrix' is the Latin word for womb, which generally means any place in which something is formed or produced. Matrices have wide applications in daily life, but are less frequently discussed.

4.2 APPLICATION OF MATRIX IN REAL LIFE

1. Matrix in Encryption
   - Used to scramble data for security purposes in encryptions, matrices are employed to encode and decode the data. There is a key which is generated using matrices which helps to encode and decode data.
   - For coding or encrypting a message, matrices and their inverse matrices are used by a programmer.
   - For communication, a message is made as a sequence of numbers in binary format, and it follows code theory for solving. Using matrices, those equations are solved.
   - Internet functions and even banks work with the transmission of sensitive and private data with these encryptions only.
2. Matrix in Computer Science
   - In the field of computer science, matrices are used in the encryption of messages. Message encryptions are used to create 3D graphic images and realistic looking motions on a 2D computer screen.
   - Eigen vector solvers are used in the calculation of algorithms that create Google Page Rankings.
   - Matrix mathematics are used by graphic software to process linear transformation to render images.
3. Matrix in Geology
   - Matrices are used in geology for taking seismic surveys.
4. Matrix in Animation
   - Matrices help to make animation more precise and perfect.

5. Matrix in Games Especially 3D
   - Matrices are used here to alter the object in 3D space.
   - 3D matrix to 2D matrix is used to convert it into different objects as per requirement.
6. Matrix in Economics and Business
   - Matrices are used to calculate the gross domestic products in economics which ultimately helps to efficiently calculate goods production.
   - Matrices are also used to study the trends of a business, shares, etc. and to help develop business models, etc.
7. Dance – Contra Dance
   - To organize complicated group dances, matrices are used.
8. Matrix in Construction
   - It is observed by everyone that some buildings appear straight, but some architects try to change the outer structure of a building, which can be done using matrices.
   - Matrices help support various historical structures.
9. Matrix in Physics
   - Matrices are essential in solving the problems using Kirchhoff's law of voltage and current.
   - Matrices help compress electronic information and play a role in storing fingerprint information.
   - Using matrices, errors in electronic transmission are identified and corrected.
   - Matrices play a vital role in calculating battery power outputs and resistor conversion of electrical energy into another useful energy.
   - Matrices are used to study electrical circuits, quantum mechanics and optics.
   - The science of optics uses matrix mathematics to account for reflection and refraction.
   - Matrix arithmetic is used to calculate the electrical properties of a circuit with voltage, amperage, resistance and the like.
   - Based on stochastic matrices, computers run Markov simulations in order to model events ranging from gambling through weather forecasting to quantum mechanics.
10. Matrix in Robotics
    - Matrices are the primary elements in the movement of robots in robotics and automation.
    - The movements of robots are programmed with the help of the calculation of matrix rows and columns.
    - The inputs for controlling robots are given based on the calculation of matrices.
11. Matrix in Medicine
    - CAT scans and MRIs use matrices in the field of medicine.

12. Other Uses of Matrices
   - Matrix calculus is used to generalize analytical notions like exponentials and derivatives to their higher dimension.
   - Matrices are the best representation method for plotting common survey things. They are used to plot graphs and are also used in statistics.
   - Matrices are used to carry out scientific study in various fields.
   - Scientists use matrices for recording the data of experiments in many organizations.
   - Matrices are used to represent real-world data about specific populations, such as the number of people who have specific traits.
   - Matrices can also be used to model projections in population growth.
   - Matrices are used to cover channels, hidden text within web pages, hidden files in plain sight, null ciphers and stenography.
   - Matrices in the form of stenography are used in a recent wireless internet connection through mobile phones known as a wireless application protocol.

13. Matrix in Cryptography
   - In the 20th century, methods of encryption and applications of mathematical theory became widespread in military usage. The military would encode messages before sending, and the recipient would decode the message in order to send safe and secure information about their military operations. Thus, the basic idea of cryptography is that information can be encoded using an encryption scheme and decoded by anyone who knows the scheme.
   - Lester Hill, a mathematics professor who taught at several US colleges and was also involved in military encryption, examined a method of encryption known as the Hill Algorithm, which uses matrix multiplication and matrix inverses. However, with the passage of time and modern computing technology, the Hill Algorithm method was not considered a secure encryption method in those days.
   - With the advancement of the computer age and Internet communication, the use of encryption, which is no longer limited to military use, has widely spread in communication and keeps the private data secure as well.
   - Lots of encryption schemes range from very simple to very complex. Modern encryption methods are more complex, often combining various steps or methods to encrypt data such as passwords, personal identification numbers, credit card numbers, social security numbers, bank account details, or corporate secrets to keep it more confidential and harder to break in this heavy usage of Internet. Some of the modern methods make use of matrices as a part of encryption and decryption process, whereas the other field of mathematics such as number theory play a vital role in modern cryptography.
   - The Greek word 'Krypto' means 'hidden'. Hence a cryptogram is a message written according to a secret code. The following describes a

method known as 'matrix multiplication' to encode and decode messages. To begin, allot a number to each letter in the alphabet (0 allocated to a blank space) as follows:

0	=	_		14	=	N
1	=	A		15	=	O
2	=	B		16	=	P
3	=	C		17	=	Q
4	=	D		18	=	R
5	=	E		19	=	S
6	=	F		20	=	T
7	=	G		21	=	U
8	=	H		22	=	V
9	=	I		23	=	W
10	=	J		24	=	X
11	=	K		25	=	Y
12	=	L		26	=	Z
13	=	M				

Then convert the message to numbers and partition it into uncoded row matrices, each having n entries, which is demonstrated in Illustration 4.1.

Illustration 4.1: Forming Uncoded Row Matrices

Write the uncoded row matrices of size 1×3 for the message MEET ME SUNDAY.

Solution: Partitioning the message into groups of three (including blank spaces, but ignoring punctuation) generates the following uncoded row matrices:

$$\begin{bmatrix} 13 & 5 & 5 \end{bmatrix} \quad \begin{bmatrix} 20 & 0 & 13 \end{bmatrix} \quad \begin{bmatrix} 5 & 0 & 19 \end{bmatrix} \quad \begin{bmatrix} 21 & 14 & 4 \end{bmatrix} \quad \begin{bmatrix} 1 & 25 & 0 \end{bmatrix}$$

$$\downarrow \downarrow \downarrow \qquad \downarrow \downarrow \downarrow \qquad \downarrow \downarrow \downarrow \qquad \downarrow \downarrow \downarrow \qquad \downarrow \downarrow \downarrow$$

M E E   T _ M   E _ S   U N D   A Y _

Note: The use of blank space is to fill out the last uncoded row matrix.

Now, to encode a message, choose an $n \times n$ invertible matrix A of our choice and multiply the uncoded row matrices by A on the right to obtain coded row matrices. This process is demonstrated in Illustration 4.2.

Illustration 4.2: Encoding a Message

Encode the message MEET ME SUNDAY by using the following invertible matrix:

$$A = \begin{bmatrix} 1 & -2 & 2 \\ -1 & 1 & 3 \\ 1 & -1 & -4 \end{bmatrix}$$

Solution: Multiply each of the uncoded row matrices found in Illustration 4.1by the given matrix A to obtain the coded row matrices as follows:

Uncoded Row Matrix	Encoding Matrix, A		Coded Row Matrix
[13 5 5]	$\begin{bmatrix} 1 & -2 & 2 \\ -1 & 1 & 3 \\ 1 & -1 & -4 \end{bmatrix}$	=	[13 −26 21]
[20 0 13]	$\begin{bmatrix} 1 & -2 & 2 \\ -1 & 1 & 3 \\ 1 & -1 & -4 \end{bmatrix}$	=	[33 −53 −12]
[5 0 19]	$\begin{bmatrix} 1 & -2 & 2 \\ -1 & 1 & 3 \\ 1 & -1 & -4 \end{bmatrix}$	=	[24 −29 19]
[21 14 4]	$\begin{bmatrix} 1 & -2 & 2 \\ -1 & 1 & 3 \\ 1 & -1 & -4 \end{bmatrix}$	=	[11 −32 68]
[1 25 0]	$\begin{bmatrix} 1 & -2 & 2 \\ -1 & 1 & 3 \\ 1 & -1 & -4 \end{bmatrix}$	=	[−24 23 77]

Thus, the sequence of coded row matrices is

$$\begin{bmatrix} 13 & -26 & 21 \end{bmatrix} \begin{bmatrix} 33 & -53 & -12 \end{bmatrix} \begin{bmatrix} 24 & -29 & 19 \end{bmatrix} \begin{bmatrix} 11 & -32 & 68 \end{bmatrix} \begin{bmatrix} -24 & 23 & 77 \end{bmatrix}$$

Removing the matrix notation finally produces the following cryptogram:

13 −26 21 33 −53 −12 24 −29 19 11 −32 68 −24 23 7

- If one does not know the encoding matrix A, decoding the cryptogram found in illustration 4.2 will be difficult. But decoding is relatively simple for an authorized receiver who knows the encoding matrix A. To retrieve the uncoded row matrices, the receiver just needs to multiply the coded row matrices by A^{-1}. In simple words, if $X = \begin{bmatrix} x_1 & x_2 & \ldots & x_n \end{bmatrix}$ is an uncoded $1 \times n$ matrix, then $Y = XA$ is the corresponding encoded matrix. By multiplying A^{-1} on the right by Y, the receiver of the encoded matrix can decode Y to obtain

$$YA^{-1} = (XA)A^{-1} = X$$

This process is demonstrated in Illustration 4.3.

Illustration 4.3: Decoding a Message

Use the inverse of the matrix $A = \begin{bmatrix} 1 & -2 & 2 \\ -1 & 1 & 3 \\ 1 & -1 & -4 \end{bmatrix}$ to decode the cryptogram

13 −26 21 33 −53 −12 24 −29 19 11 −32 68 −24 23 7

Solution: The inverse of the matrix $A = \begin{bmatrix} 1 & -2 & 2 \\ -1 & 1 & 3 \\ 1 & -1 & -4 \end{bmatrix}$ is $A^{-1} = \begin{bmatrix} -1 & -10 & -8 \\ -1 & -6 & -5 \\ 0 & -1 & -1 \end{bmatrix}$.

To decode the message, partition the message into groups of three to form the coded row matrices

$$\begin{bmatrix} 13 & -26 & 21 \end{bmatrix} \begin{bmatrix} 33 & -53 & -12 \end{bmatrix} \begin{bmatrix} 24 & -29 & 19 \end{bmatrix} \begin{bmatrix} 11 & -32 & 68 \end{bmatrix} \begin{bmatrix} -24 & 23 & 77 \end{bmatrix}$$

Multiply each coded row matrix by A^{-1} on the right to obtain the decoded row matrices.

Coded Row Matrix	Decoding Matrix A^{-1}		Decoded Row Matrix
[13 −26 21]	$\begin{bmatrix} -1 & -10 & -8 \\ -1 & -6 & -5 \\ 0 & -1 & -1 \end{bmatrix}$	=	[13 5 5]
[33 −53 −12]	$\begin{bmatrix} -1 & -10 & -8 \\ -1 & -6 & -5 \\ 0 & -1 & -1 \end{bmatrix}$	=	[20 0 13]
[24 −29 19]	$\begin{bmatrix} -1 & -10 & -8 \\ -1 & -6 & -5 \\ 0 & -1 & -1 \end{bmatrix}$	=	[5 0 19]
[11 −32 68]	$\begin{bmatrix} -1 & -10 & -8 \\ -1 & -6 & -5 \\ 0 & -1 & -1 \end{bmatrix}$	=	[21 14 4]
[−24 23 77]	$\begin{bmatrix} -1 & -10 & -8 \\ -1 & -6 & -5 \\ 0 & -1 & -1 \end{bmatrix}$	=	[1 25 0]

The sequence of the decoded row matrices is

$$\begin{bmatrix} 13 & 5 & 5 \end{bmatrix} \begin{bmatrix} 20 & 0 & 13 \end{bmatrix} \begin{bmatrix} 5 & 0 & 19 \end{bmatrix} \begin{bmatrix} 21 & 14 & 4 \end{bmatrix} \begin{bmatrix} 1 & 25 & 0 \end{bmatrix}$$

Thus, the message is

13 5 5 20 0 13   5   0 19 21 14 4 1 25 0

↓ ↓ ↓ ↓ ↓ ↓   ↓ ↓ ↓ ↓   ↓ ↓ ↓ ↓ ↓

M E E T - _ M   E _ S U   N D A Y _

- Matrix encryption is just one of many schemes. Every year, the National Security Agency, the military force and private corporations hire hundreds of people to launch new schemes and decode existing ones.

4.3   INTRODUCTION TO APPLICATION OF DETERMINANT

Historically, determinants were used long before matrices. Determinants were first used by Chinese scholars in the 3rd century BC and were used in the Chinese mathematics textbook *The Nine Chapters on the Mathematical Art,* which is much earlier than the use of matrices themselves (AD 1850). The applications and uses of determinants are discussed in detail in the next section.

4.4   APPLICATION OF DETERMINANT

- One can find the use of the determinant whenever one has to represent something in the form of a matrix, by understanding the fact that all images that one sees are represented in some form of matrix in a computer.
- Extensive use of matrices is required for 3D representations, graphics, etc., where one may need the use of determinants. They are mainly used in the field of science and engineering. Hence, the engineers of various streams may require them in the process of their practice and use in applications.
- Determinants are used to
  - i. Determine whether the matrix has an inverse or not.
  - ii. Determine the characteristic polynomial of a matrix and hence its eigen values.
  - iii. Find the adjoint of a matrix and using it to find the inverse.
  - iv. Solve a system of linear equations in Cramer's Rule.
  - v. Find area, volume and the equations of lines and planes.
- We are aware of the methods to use determinants in points i, ii, iii and iv from the previous chapters. Let's see how determinants are used in point v.
- Area of a triangle in XY plane.

The area of triangle with vertices (x_1, y_1), (x_2, y_2) and (x_3, y_3) is

$$\text{Area} = \pm\frac{1}{2}\begin{vmatrix} x_1 & y_1 & 1 \\ x_2 & y_2 & 1 \\ x_3 & y_3 & 1 \end{vmatrix}$$

where the sign $(\pm)$ is chosen to assign a positive area.

- Test for collinear points in the XY plane

  Three points (x_1, y_1), (x_2, y_2) and (x_3, y_3) are collinear if and only if

$$\begin{vmatrix} x_1 & y_1 & 1 \\ x_2 & y_2 & 1 \\ x_3 & y_3 & 1 \end{vmatrix} = 0$$

- Two-point form of the equation of a line

  An equation of the line passing through the distinct points (x_1, y_1) and (x_2, y_2) is given by

$$\begin{vmatrix} x & y & 1 \\ x_1 & y_1 & 1 \\ x_2 & y_2 & 1 \end{vmatrix} = 0$$

- Three-point form of the equation of a plane

  An equation of the plane passing through the distinct points (x_1, y_1, z_1), (x_2, y_2, z_2) and (x_3, y_3, z_3) is given by

$$\begin{vmatrix} x & y & z & 1 \\ x_1 & y_1 & z_1 & 1 \\ x_2 & y_2 & z_2 & 1 \\ x_3 & y_3 & z_3 & 1 \end{vmatrix} = 0$$

- Test for coplanar points in space

  Four points (x_1, y_1, z_1), (x_2, y_2, z_2), (x_3, y_3, z_3) and (x_4, y_4, z_4) are coplanar if and only if

$$\begin{vmatrix} x_1 & y_1 & z_1 & 1 \\ x_2 & y_2 & z_2 & 1 \\ x_3 & y_3 & z_3 & 1 \\ x_4 & y_4 & z_4 & 1 \end{vmatrix} = 0$$

- Volume of a tetrahedron

  The volume of a tetrahedron with vertices (x_1, y_1, z_1), (x_2, y_2, z_2), (x_3, y_3, z_3) and (x_4, y_4, z_4) is

$$\text{Volume} = \pm\frac{1}{6}\begin{vmatrix} x_1 & y_1 & z_1 & 1 \\ x_2 & y_2 & z_2 & 1 \\ x_3 & y_3 & z_3 & 1 \\ x_4 & y_4 & z_4 & 1 \end{vmatrix}$$

where the sign ($\pm$) is chosen to assign a positive volume.
- To determine the equation of the orbit of a planet, an astronomer can find the coordinates of the planet along its orbit at five different points (x_i, y_i), where i = 1, 2, 3, 4, 5, and then use the determinant

$$\begin{vmatrix} x^2 & xy & y^2 & x & y & 1 \\ x_1^2 & x_1y_1 & y_1^2 & x_1 & y_1 & 1 \\ x_2^2 & x_2y_2 & y_2^2 & x_2 & y_2 & 1 \\ x_3^2 & x_3y_3 & y_3^2 & x_3 & y_3 & 1 \\ x_4^2 & x_4y_4 & y_4^2 & x_4 & y_4 & 1 \\ x_5^2 & x_5y_5 & y_5^2 & x_5 & y_5 & 1 \end{vmatrix}$$

MULTIPLE-CHOICE QUESTIONS

1. Matrix means any place in which something is _____.
   (a) Destroyed
   (b) Formed
   (c) Produced
   (d) Both (b) and (c)
2. For coding or encrypting a message, matrices and their _____ matrices are used by a programmer.
   (a) Inverse
   (b) Skew symmetric
   (c) Transpose
   (d) None of these
3. Matrix is the _____ element in the movement of robots in robotics and automation.
   (a) Primary
   (b) Secondary
   (c) Tertiary
   (d) None of these
4. The determinant is the _____ of the matrix.
   (a) Size
   (b) Length
   (c) Both (a) and (b)
   (d) None of these
5. Determinants are used to find _____.
   (a) Area of triangle

(b) Volume of tetrahedron
(c) Equation of line and plane
(d) All of the above

ANSWERS TO MULTIPLE-CHOICE QUESTIONS

Answer 1: (d)
Answer 2: (a)
Answer 3: (a)
Answer 4: (c)
Answer 5: (d)

Bibliography

Aparna, M. 2016. "Application of Matrix Mathematics." *International Education and Research Journal* 2 (12):80–82. ierj.in/journal/index.php/ierj/article/view/601

egyankosh.ac.in/bitstream/123456789/20409/1/Unit-10.pdf

homepage.ntu.edu.tw/~jryanwang/course/Mathematics%20for%20Management%20(undergraduate%20level)/Applications%20in%20Ch2.pdf

homepage.ntu.edu.tw/~jryanwang/course/Mathematics%20for%20Management%20(undergraduate%20level)/Applications%20in%20Ch3.pdf

http://www.pbte.edu.pk/text%20books/dae/math_113/Chapter_09.pdf

https://gradeup.co/matrices-and-determinants-iit-jee-notes-i

https://math.stackexchange.com/questions/553981/what-are-the-applications-of-matrices-in-real-world

https://open.umn.edu/opentextbooks/textbooks/fundamentals-of-matrix-algebra

https://users.soe.ucsc.edu/~hongwang/AMS10/Week_8/sec5.1-5.3.pdf

https://www.academia.edu/30903815/Eigenvalues_and_Eigenvectors_Matrices_Eigenvalues_and_Eigenvectors_Matrices_Eigenvalues_and_Eigenvectors

https://www.embibe.com/exams/where-are-matrices-used-in-daily-life/undefined

https://www.hec.ca/en/cams/help/topics/Matrix_determinants.pdf

https://www.math.ust.hk/~machas/matrix-algebra-for-engineers.pdf

https://www.saddleback.edu/faculty/fgonzalez/Spring_2010/Math_2/csg_ch08.pdf

https://www.slideshare.net/moneebakhtar50/application-of-matrices-in-real-life

https://www.vmi.edu/media/content-assets/documents/academics/appliedmath/Fundamentals-of-Matrix-Algebra-3rd-Edition.pdf

mathsguideonline.weebly.com/blog/real-life-application-of-matrices-some-examples

Saff, E. B., and Snider, A. D. 2015. *Fundamentals of Matrix Analysis with Applications.* John Wiley & Sons: Hoboken, New Jersey.

vcp.med.harvard.edu/papers/matrices-1.pdf

www.cengage.com/resource_uploads/downloads/1439049254_242720.pdf

www.math.utep.edu/Faculty/cmmundy/Math%202301/0495292974_CHPT_03.pdf

Index

For Product Safety Concerns and Information please contact our EU
representative GPSR@taylorandfrancis.com
Taylor & Francis Verlag GmbH, Kaufingerstraße 24, 80331 München, Germany